ALANNA COLLEN is a science writer, with both bachelor's and master's degrees in biology from Imperial College London, and a PhD in evolutionary biology from University College London and the Zoological Society of London. She is a well-travelled zoologist, an expert in bat echolocation, and an accidental collector of tropical diseases. During her scientific career, Alanna has written for the *Sunday Times Magazine*, as well as about wildlife for ARKive.org. She has appeared on both radio and television, including BBC Radio 4's *The Tribes of Science* and *Saturday Live*, and BBC One's adventure-wildlife show *Lost Land of the Volcano*. She lives in Bedfordshire with her husband.

10% Human

How Your Body's Microbes
Hold the Key to Health
and Happiness

ALANNA COLLEN

WILLIAM
COLLINS

William Collins
An imprint of HarperCollins*Publishers*
1 London Bridge Street
London SE1 9GF
WilliamCollinsBooks.com

First published in Great Britain by William Collins in 2015

This William Collins paperback edition published 2016

1

A catalogue record for this book
is available from the British Library

ISBN 978-0-00-758405-5

Line drawings by Mat Taylor
Set in Minion by Birdy Book Design

Printed and bound in Great Britain by Clays Ltd, St Ives plc

MIX
Paper from
responsible sources
FSC™ C007454
www.fsc.org

FSC™ is a non-profit international organisation established to promote the
responsible management of the world's forests. Products carrying the FSC
label are independently certified to assure consumers that they come from
forests that are managed to meet the social, economic and ecological needs
of present and future generations, and other controlled sources.

Find out more about HarperCollins and the environment at
www.harpercollins.co.uk

For Ben and his microbes.
My favourite superorganism.

At the heart of science is an essential balance between two seemingly contradictory attitudes – an openness to new ideas, no matter how bizarre or counterintuitive they may be, and the most ruthless sceptical scrutiny of all ideas, old and new. This is how deep truths are winnowed from deep nonsense.

CARL SAGAN

CONTENTS

PROLOGUE

Being Cured

As I walked back through the forest that night in the summer of 2005, with twenty bats in cotton bags hanging around my neck and all manner of insect life dashing for the light of my head torch, I realised my ankles were itching. I had my repellent-soaked trousers tucked into my leech socks, with another pair underneath for good measure. The humidity and drenching sweat, the muddy trails, my fear of tigers, and the mosquitoes were enough to contend with as I made my rounds, collecting bats out of traps in the darkness of the rainforest. But something had got through the barrier of fabric and chemicals protecting my skin. Something itchy.

At twenty-two, I spent what turned out to be a life-changing three months living in the heart of Krau Wildlife Reserve in peninsular Malaysia. During my biology degree, I had become fascinated by bats, and when the opportunity came up to work as a field assistant to a British bat scientist, I signed up immediately. Encounters with leaf monkeys, gibbons and an extraordinary diversity of bats made the challenges of sleeping in a hammock and washing in a river populated by monitor lizards seem worthwhile. But, as I was to discover, the trials of life in a tropical forest can live on far beyond the experience itself.

Back at base camp, in a clearing next to the river, I peeled back the layers to reveal the source of my discomfort: not leeches, but ticks.

Perhaps fifty or so, some embedded in my skin, others crawling up my legs. I brushed the loose ones off, and turned back to the bats, measuring and recording scientific data about them as quickly as I could. Later, with the bats released, and the forest pitch-black and buzzing with cicadas, I zipped myself into my cocoon-like hammock, and, with a pair of tweezers, under the light of my head torch, I removed every last tick.

A few months later, at home in London, the tropical infection introduced to me by the ticks took hold. My body seized up and my toe bone swelled. Weird symptoms came and went, as did various blood tests and hospital specialists. My life would be put on hold for weeks or months at a time as bouts of pain, fatigue and confusion gripped me without warning, then released me again as if nothing had happened. By the time I was diagnosed many years later, the infection was entrenched, and I was given a course of antibiotics long and intense enough to cure a herd of cattle. At last, I was going to be myself again.

But, unexpectedly, the story did not end there. I was cured, but not just of the tick-borne infection. Instead, it seemed I had been cured as if I were a piece of meat. The antibiotics had worked their magic, but I began to suffer new symptoms, as varied as before. My skin was raw, my digestive system was choosy, and I was prone to picking up every infection going. I had a suspicion that the antibiotics I had taken had not only eradicated the bacteria that plagued me, but also those that belonged in me. I felt like I had become inhospitable to microbes, and I learnt just how much I needed the 100 trillion friendly little creatures who had, until recently, called my body their home.

You are just 10 per cent human.

For every one of the cells that make up the vessel that you call your body, there are nine impostor cells hitching a ride. You are not just flesh and blood, muscle and bone, brain and skin, but also bacteria and fungi. You are more 'them' than you are 'you'. Your gut alone hosts 100 trillion of them, like a coral reef growing on the rugged seabed that is your intestine. Around 4,000 different species carve out their own little niches, nestled among folds that give your 1.5-metres-long colon the surface area of a double bed. Over your lifetime, you will play host to bugs the equivalent weight of five African elephants. Your skin is crawling with them. There are more on your fingertip than there are people in Britain.

Disgusting, isn't it? We are surely too sophisticated, too hygienic, too *evolved* to be colonised in this way. Shouldn't we have shunned microbes, like we did fur and tails, when we left the forests? Doesn't modern medicine have the tools to help us evict them so that we can live cleaner, more healthy, independent lives? Since the body's microbial habitat was first discovered we have tolerated it, as it seemed to do us no harm. But unlike the coral reefs, or the rainforests, we have not thought to protect it, let alone to cherish it.

As an evolutionary biologist, I am trained to look for the advantage, the *meaning*, in the anatomy and behaviour of an organism. Usually, characteristics and interactions that are truly detrimental are either fought against, or lost in evolutionary time. That set me thinking: our 100 trillion microbes could not call us home if they brought nothing to the party. Our immune systems fight off germs and cure us of infections, so why would they tolerate being invaded in this way? Having subjected my own invaders, both good and bad, to months of chemical warfare, I wanted to know more about the collateral damage I had caused.

As it turned out, I was asking this question at exactly the right time. After decades of slow-paced scientific attempts at learning more about the body's microbes by culturing them on Petri dishes, technology had finally caught up with our curiosity. Most of the microbes living inside us die when they are exposed to oxygen, because they are adapted to an oxygen-free existence deep in our guts. Growing them outside the body is difficult, and experimenting with them is even harder.

But, in the wake of the seminal Human Genome Project, in which every human gene was decoded, scientists are now capable of sequencing massive quantities of DNA extremely quickly and cheaply. Even our dead microbes, expelled from the body in the stool, could now be identified because their DNA remained intact. We had thought our microbes didn't matter, but science is beginning to reveal a different story. A story in which our lives are intertwined with those of our hitchhikers, where our microbes run our bodies, and becoming a healthy human is impossible without them.

My own health troubles were the tip of the iceberg. I learnt of the emerging scientific evidence that disruptions to the body's microbes were behind gastrointestinal disorders, allergies, autoimmune diseases, and even obesity. And it wasn't just physical health that could be affected, but mental health as well, from anxiety and depression to obsessive–compulsive disorder (OCD) and autism. Many of the illnesses we accept as part of life were not, it seemed, down to flaws in our genes, or our bodies letting us down, but were instead newly emerging conditions brought on by our failure to cherish the long-held extension to our own human cells: our microbes.

Through my research, I hoped not only to discover what damage the antibiotics I had taken had done to my microbial colony, but how it had made me unwell, and what I could do to restore the balance of microbes I had harboured before the night of the tick bites, eight years earlier. To learn more, I signed up to take the ultimate step in self-discovery: DNA sequencing. But rather than sequence my own genes, I would have the genes of my personal colony of microbes – my microbiome – sequenced. By knowing which species and strains of bacteria I contained, I would have a starting point for self-improvement. Using the latest understanding of what *should* be living in me, I might be able to judge just how much damage I had done, and attempt to make amends. I used a citizen-science programme, the American Gut Project, based at the laboratory of Professor Rob Knight at the University of Colorado, Boulder. Available to anyone around the world for a donation, the AGP sequences samples of microbes from the human body to learn more about the species we

4

harbour and their impact on our health. By sending a stool sample containing the microbes from my own gut, I received a snapshot of the ecosystem that called my body home.

After years of antibiotics, I was relieved to find I had *any* bacteria living in me at all. It was pleasing to see that the groups I harboured were at least broadly similar to those in other American Gut Project participants, and not the microbial equivalent of mutant creatures eking out a living on a toxic wasteland. But, perhaps predictably, the diversity of my bacteria seemed to have taken a beating. At the highest level of the taxonomic hierarchy, the diversity was relatively low, looking a bit bipartisan compared with the guts of other people. Over 97 per cent of my bacteria belonged to the two major bacterial groups, compared with around 90 per cent making up these two groups in the average participant. Perhaps the antibiotics I had taken had killed off some of the less abundant species, leaving me with only the hardy survivors. I was intrigued to know whether this loss might be related to any of my more recent health troubles.

But, just as comparing a tropical rainforest and an oak woodland by looking at the proportion of trees to shrubs, or birds to mammals, reveals little about how both ecosystems function, comparing my bacteria at such a broad scale may not tell me all that much about the health of my inner community. At the other end of the taxonomic hierarchy were the genera and species that I contained. What could the identities of the bacteria that had either clung on throughout my treatment, or returned since it ended, reveal about my current state of health? Or perhaps more pertinently, what did the *absence* of species that might have fallen victim to the chemical warfare I had unleashed on them, mean for me now?

As I embarked on learning more about *us* – myself and my microbes – I resolved to put what I learnt into practice. I wanted to get back on their good side, and I knew I needed to make changes to my life to restore a colony that would work in harmony with my human cells. If my most recent symptoms were stemming from the collateral damage I had inadvertently inflicted upon my microbiota, perhaps I could reverse it and rid myself of the allergies, the skin problems and the near-constant infections? My concern wasn't just for myself, but for the children I hoped to have in the coming years. As I would pass on not only my genes, but also my microbes, I wanted to be sure I had something worth giving.

I resolved to put my microbes first, altering my diet to better suit their needs. I planned to have a second sample sequenced after my lifestyle changes had had a chance to take effect, in the hope that my efforts might be evident from the change in the diversity and balance of the species I play host to. Most of all, I hoped that my investment in them would pay dividends, by unlocking the door to better health and happiness.

INTRODUCTION

The Other 90%

In May 2000, just weeks before the announcement of the first draft of the human genome, a notebook began circulating among the scientists sitting at the bar in Cold Spring Harbor Laboratory in New York State. Excitement was building about the next phase of the Human Genome Project, in which the DNA sequence would be split into its functional parts – genes. The notebook contained a sweepstake: the guesses of the best-informed group of people on the planet concerning one intriguing question: How many genes does it take to build a human?

Senior research scientist Lee Rowen, who was leading a group working on decoding chromosomes 14 and 15, sipped her beer as she pondered the question. Genes make proteins, the building blocks of life, and the sheer complexity of humans made it seem probable that the number would be high. Higher than the mouse, surely, which was known to have 23,000 genes. Probably also higher than the wheat plant, with 26,000 genes. And, no doubt, far higher than 'The Worm', a favourite laboratory species of developmental biologists, with its 20,500 genes.

Despite guesses averaging over 55,000 genes, and topping 150,000, Rowen's understanding of the field meant she was inclined to go low. She placed a bet of 41,440 that year, and followed it up a year later with a second bet of just 25,947 genes. In 2003, with the true gene number only just emerging from the nearly finished sequence, Rowen was awarded the prize. Her entry was the lowest of all 165 bets, and the latest gene count had just dropped even lower than any scientist had ever predicted.

With just shy of 21,000 genes, the human genome is hardly bigger than that of The Worm (*C. elegans*). It is half the size of the rice plant, and even the humble water flea outstrips it, with 31,000 genes. None of these species can talk, create, or think intelligent thoughts. You might think, as the scientists entering the Genesweep pool did, that humans would have a great many more genes than grasses and worms and fleas. After all, genes make proteins, and proteins make bodies. Surely a body as complex and sophisticated as a human's would need more proteins, and therefore more genes, than a worm's?

But these 21,000 genes are not the only genes that run your body. We do not live alone. Each of us is a superorganism; a collective of species, living side-by-side and cooperatively running the body that sustains us all. Our own cells, though far larger in volume and weight, are outnumbered ten to one by the cells of the microbes that live in and on us. These 100 trillion microbes – known as the microbiota – are mostly bacteria: microscopic beings made of just a single cell each. Alongside the bacteria are other microbes – viruses, fungi and archaea. Viruses are so small and simple that they challenge our ideas of what even constitutes 'life'. They depend entirely on the cells of other creatures to replicate themselves. The fungi that live on us are often yeasts; more complex than bacteria, but still small, single-celled organisms. The archaea are a group that appear to be similar to bacteria, but they are as different evolutionarily as bacteria are from plants or animals. Together, the microbes living on the human body contain 4.4 *million* genes – this is the microbiome: the collective genomes of the microbiota. These genes collaborate in running our bodies alongside our 21,000 human genes. By that count, you are just half a per cent human.

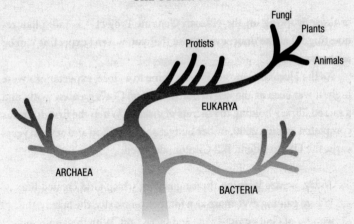

*A simplified tree of life, showing the three domains
and four kingdoms of Domain Eukarya.*

We now know that the human genome generates its complexity not only in the number of genes it contains, but also through the many combinations of proteins these genes are able to make. We, and other animals, are able to extract more functions from our genomes than they appear to encode at first glance. But the genes of our microbes add even more complexity to the mix, providing services to the human body that are more quickly evolved and more easily provided by these simple organisms.

Until recently, studying these microbes relied on being able to culture them on Petri dishes filled with broths of blood, bone marrow, or sugars, suspended in jelly. It's a difficult task: most of the species living in the human gut die on exposure to oxygen – they simply haven't evolved to tolerate it. What's more, growing microbes on these plates means guessing what nutrients, temperature and gases they might need to survive, and failing to figure this out means failing to learn more about a species. Culturing microbes is the equivalent of checking who's turned up for class by running down a register – if you don't call someone's name, you won't know if they are there. Today's technology – the DNA sequencing made so fast and cheap by the efforts of those working on the Human Genome Project – is more like requesting ID at the door; even those that you weren't expecting can be accounted for.

As the Human Genome Project came to a close, expectations were high. It was seen as the key to our humanity, God's greatest work, and a sacred library holding the secrets of disease. When the first draft was completed in June 2000, under budget at $2.7 billion and several years early, the US President, Bill Clinton, declared:

> Today, we are learning the language in which God created life. We are gaining ever more awe for the complexity, the beauty, the wonder of God's most divine and sacred gift. With this profound new knowledge, humankind is on the verge of gaining immense new power to heal. Genome science will have a real impact on all our lives – and even more, on the lives of our children. It will revolutionise the diagnosis, prevention and treatment of most, if not all, human diseases.

But in the years that followed, science journalists the world over began expressing their disappointment in the contribution that knowledge of our complete DNA sequence had made to medicine. Although decoding our own instruction book is an irrefutable achievement that has made a difference to treatments for several important illnesses, it has not revealed as much as we expected about the causes of many common diseases. Searching for genetic differences in common to people with a particular disease did not throw up straightforward links for as many conditions as had been expected. Often, conditions were weakly linked to tens or hundreds of gene variants, but rarely was it the case that possessing a given gene variant would lead directly to a given disease.

What we failed to appreciate at the turn of the century was that those 21,000 genes of ours are not the full story. The DNA-sequencing technology invented during the Human Genome Project enabled another major genome-sequencing programme, but one that received far less media attention: the Human Microbiome Project. Rather than looking at the genome of our own species, the HMP was set up to use the genomes of the microbes that live on the human body – the microbiome – to identify which species are present.

No longer would a reliance on Petri dishes and an over-abundance of oxygen hold back research into our cohabiters. With a budget of $170 million and a five-year programme of DNA sequencing, the HMP was to read thousands of times as much DNA as the HGP, from microbes living in eighteen different habitats on the human body. It was to be a far more comprehensive survey of the genes that make a person, both human and microbial. At the conclusion of the Human Microbiome Project's first phase of research in 2012, not one world leader made a triumphant statement, and only a handful of newspapers featured the story. But the HMP would go on to reveal more about what it means to be human today than our own genome ever has.

Since life began, species have exploited one another, and microbes have proved themselves to be particularly efficient at making a living in the oddest of places. At their microscopic size, the body of another organism, particularly a macro-scale backboned creature like a human, represents not just a single niche, but an entire world of habitats, ecosystems and opportunities. As variable and dynamic as our spinning planet, the human body has a chemical climate that waxes and wanes with hormonal tides, and complex landscapes that shift with advancing age. For microbes, this is Eden.

We have been co-evolving side-by-side with microbes since long before we were humans. Before our ancestors were mammals even. Each animal body, from the tiniest fruit fly to the largest whale, is yet another world for microbes. Despite the negative billing many of them get as disease-causing germs, playing host to a population of these miniature life-forms can be extremely rewarding.

The Hawaiian bobtail squid – as big-eyed and colourful as any Pixar character – has diminished a major threat to its life by inviting just one species of bioluminescent bacterium to live in a special cavity in its underbelly. Here, in this light organ, the bacteria, known as *Aliivibrio fischeri*, convert food into light, so that viewed from below, the squid glows. This obscures its silhouette against the moonlit ocean surface, camouflaging it from predators approaching from beneath. The squid owes this protection to its bacterial inhabitants, and they owe the squid for their home.

While housing a microbial light source might seem a particularly inventive way to increase one's life chances, squid are far from the only animal species who owe their lives to their body's microbes. Strategies for living are many and varied, and cooperation with microbes has been a driving force of the evolutionary game since living beings with more than one cell first evolved, 1.2 billion years ago.

The more cells an organism is made of, the more microbes can live on it. Indeed, large animals such as cattle are well known for their bacterial hospitality. Cows eat grass, yet using their own genes they can extract very little nutrition from this fibrous diet. They would need specialist proteins, called enzymes, that can break down the tough molecules making the cell walls of the grass. Evolving the genes that make these enzymes could take millennia, as it relies on random mutations in the DNA code that can only happen with each passing generation of cows.

A quicker way to acquire the ability to get at the nutrients locked away in grass is to outsource the task to the specialists: microbes. The four chambers of the cow's stomach house populations of plant-fibre-busting microbes numbering in their trillions, and the cud – a ball of solid plant fibre – travels back and forth between the mechanical grinding of the cow's mouth and chemical breakdown by the enzymes produced by microbes living in the gut. Acquiring the genes to do this is quick and easy for microbes, as their generation times, and therefore opportunities for mutations and evolution, are often less than a day.

If bobtail squid and cows can both benefit from teaming up with microbes, is it possible that we humans do as well? We may not eat grass and have a four-chambered stomach, but we do have our own specialisations. Our stomachs are small and simple, there just to mix the food up, throw in some enzymes for digestion, and add a bit of acid to kill unwelcome bugs. But travel on, through the small intestine, where food is broken down by yet more enzymes and absorbed into the blood through the carpet of finger-like projections that give it the surface area of a tennis court, and you reach a cul-de-sac, more of a tennis ball than a tennis court, that marks the beginning of the large intestine. This pouch-like patch, at the lower right corner of your torso, is called the caecum, and it is the heart of the human body's microbial community.

Dangling from the caecum is an organ that has a reputation for being there simply to cause pain and infection: the appendix. Its full title – the vermiform appendix – refers to its worm-like appearance, but it could equally be compared to a maggot or a snake. Appendices vary in length from a diminutive 2 cm to a distinctly stringy 25 cm, and, rarely, a person may even have two of them, or not one at all. If popular opinion is to be believed, we would be better off without one at all, since for over one hundred years they have been said to have no function whatsoever. In fact, the man who finally put the anatomy of animals into an elegant evolutionary framework is apparently responsible for this persistent myth. Charles Darwin, in *The Descent of Man*, a follow-up to *On the Origin of Species*, included the appendix in a discussion of 'rudimentary' organs. Having compared it with the larger appendices of many other animals, Darwin felt that the appendix was a vestige, steadily withering away as humans changed their diets.

With little to indicate otherwise, the vestigial status of the appendix was barely questioned for the next 100 years, and the perception of its uselessness is only enhanced by its tendency to cause a nuisance. So pointless has the medical establishment assumed it to be, that by the 1950s, removing it became one of the most common surgical procedures carried out in the developed world. An appendectomy was even often tacked on as a bonus during other abdominal surgery. At one point, a man stood a one in eight chance of having his appendix removed during his lifetime, and for a woman, the odds were one in four. About 5–10 per cent of people get appendicitis at some stage in their life, usually in the decades before they have children. Untreated, nearly half of these people would die.

This presents a conundrum. If appendicitis were a naturally occurring disease, frequently causing death at a young age, the appendix would be quickly eliminated by natural selection. Those with appendices large enough to become infected would die, most often before reproducing, and would therefore fail to pass on their appendix-forming genes. Over time, fewer and fewer people would have an appendix, and eventually it would be lost. Natural selection would have preferred those without one.

Darwin's assumption that it was a relic of our pasts might have carried some weight, were it not for the often fatal consequences of possessing one. There are two explanations, therefore, for the persistence of the appendix, and they are not mutually exclusive. The first is that appendicitis is a modern phenomenon, brought on by some environmental change. Thus, even a pointless organ could have persisted in the past simply by keeping out of trouble. The other is that the appendix, far from being a malign vestige of our evolutionary past, actually has health benefits that outweigh its dark side, making its presence worthwhile despite the risk of appendicitis. That is, natural selection prefers those of us who possess one. The question is, why?

The answer lies in its contents. The appendix, which averages about 8 cm in length and a centimetre across, forms a tube, protected from the flow of mostly digested food passing its entrance. But rather than being a withered strand of flesh, it is packed full of specialised immune cells and molecules. They are not inert, but rather an integral part of the immune system, protecting, cultivating and communicating with a collective of microbes. Inside, these microbes form a 'biofilm' – a layer of individuals that support one another and exclude bacteria that might cause harm. The appendix, far from being functionless, appears to be a safe-house that the human body has provided for its microbial inhabitants.

Like a nest egg stashed away for a rainy day, this microbial stockpile comes in handy at times of strife. After an episode of food poisoning or a gastrointestinal infection, the gut can be repopulated with its normal inhabitants, which have been lurking in the appendix. It might seem like an excessive bodily insurance policy, but it is only in recent decades that gut infections such as dysentery, cholera and giardiasis have been all but eliminated in the Western world. Public sanitation measures, including sewerage systems and water-treatment plants, have prevented such illnesses in developed countries, but globally, one in five of all childhood deaths are still caused by infectious diarrhoea. For those who do not succumb, possession of an appendix likely hastens their recovery. It is only in a context of relatively good health that we have come to believe that the appendix has no function. Indeed, the negative consequences of undergoing an appendectomy have been masked by the modern, sanitised lifestyle.

As it turns out, appendicitis *is* a modern phenomenon. In Darwin's day, it was extremely rare, causing very few deaths, so we can perhaps forgive him for thinking the appendix was merely one of evolution's leftovers, neither harming nor helping us. Appendicitis became common in the late nineteenth century, with cases in one British hospital shooting up from a stable rate of three or four people per year prior to 1890, to 113 cases per year by 1918; a rise mirrored throughout the industrialised world. Diagnosis had never been a problem – the cramping pain followed by a quick autopsy if the patient didn't make it revealed the cause of death even before appendicitis became as common as it is now.

Many explanations were put forward to explain it, from increased meat, butter and sugar consumption, to blocked sinuses and rotting teeth. At that time, consensus opinion alighted on a reduction in fibre in our diets as the ultimate cause, but hypotheses still abound, including one that blames the rise on improved water sanitation and the hygienic conditions it brings – the very development that appeared to render the appendix almost impotent. Whatever the ultimate cause, by the Second World War our collective memory had been purged of the rise in appendicitis cases, leaving us with the impression that it is an expected, though unwelcome, feature of normal life.

In fact, even in the modern, developed world, keeping hold of the appendix at least until adulthood can prove to be beneficial, protecting us from recurring gastrointestinal infections, immune dysfunction, blood cancer, some autoimmune diseases and even heart attacks. Somehow, its role as a sanctuary of microbial life brings these benefits.

That the appendix is far from pointless tells us something bigger: our microbes matter to our bodies. It seems they are not just hitching a ride, but providing a service important enough that our guts have evolved an asylum just to keep them safe. The question is, who is there, and what exactly do they do for us?

Although we have known for several decades that our bodies' microbes offer us a few perks, like synthesising some essential vitamins, and breaking down tough plant fibres, the degree of interaction between our cells and theirs wasn't realised until relatively recently. In the late 1990s, using the tools of molecular biology, microbiologists took a great leap into discovering more about our strange relationship with our microbiotas.

New DNA-sequencing technology can tell us which microbes are present, and allows us to place them within the tree of life. With each step down this hierarchy, from domain to kingdom then phylum through class, order and family, and on to genus, species and strain, individuals are more and more closely related to one another. Working from the bottom up, we humans (genus *Homo* and species *sapiens*) are great apes (family Hominidae), which sit alongside monkeys and others within the primates (order Primates). All of us primates belong with our fellow furry milk-drinkers, as a member of the mammals (class Mammalia), who then fit within a group containing animals with a spinal cord (phylum Chordata), and finally, amongst *all* animals, spinal cord or otherwise (think of our squid, for example), in kingdom Animalia, and domain Eukarya. Bacteria and other microbes (except the category-defying viruses) find their place on the other great branches of the tree of life, belonging not to kingdom Animalia, but to their own unique kingdoms in separate domains.

Sequencing allows different species to be identified and placed within the hierarchy of the tree of life. One particularly useful segment of DNA, the 16S rRNA gene, acts as a kind of barcode for bacteria, providing a quick ID without the need to sequence an entire bacterial genome. The more similar the codes of the 16S rRNA genes, the more closely related the species, and the more twigs and branches of the tree of life they share.

DNA sequencing, though, is not the only tool at our disposal when it comes to answering questions about our microbes, especially regarding what they do. For these mysteries, we often turn to mice. In particular, 'germ-free' mice. The first generations of these laboratory staples were born by Caesarean section and kept in isolation chambers, preventing them from ever becoming colonised with microbes, either beneficial or harmful ones. From then on, most germ-free mice are simply born in isolation to germ-free mothers, sustaining a sterile line of rodents untouched by microbes. Even their food and bedding is irradiated and packed in sterile containers to prevent any contamination of the mice. Transferring mice between their bubble-like cages is quite an operation, involving vacuums and anti-microbial chemicals.

By comparing germ-free mice with 'conventional' mice, which have their full complement of microbes, researchers are able to test the exact effects of having a microbiota. They can even colonise germ-free mice with a single species of bacterium, or a small set of species, to see precisely how each strain contributes to the biology of a mouse. From studying these 'gnotobiotic' ('known life') mice, we get an inkling of what microbes do in us as well. Of course, they are not the same as humans, and sometimes results of experiments in mice are wildly different from results in humans, but they are a fantastically useful research tool and very often provide crucial leads. Without rodent models, medical science would progress at a millionth of the speed.

It was by using germ-free mice that the commander-in-chief of microbiome science, Professor Jeffrey Gordon from the University of Washington in St Louis, Missouri, discovered a remarkable indication of just how fundamental the microbiota are to the running of a healthy body. He compared the guts of germ-free mice with those of conventional mice, and he learnt that under the direction of bacteria the mouse cells lining the gut wall were releasing a molecule that 'fed' the microbes, encouraging them to take up residence. The presence of a microbiota not only changes the chemistry of the gut, but also its morphology. The finger-like projections grow longer at the insistence of microbes, making the surface area large enough to capture the energy it needs from food. It's been estimated that rats would need about 30 per cent more to eat if it weren't for their microbes.

It is not only microbes that are benefiting from sharing our bodies, but us too. Our relationship with them is not just one of tolerance, but encouragement. This realisation, combined with the technical power of DNA sequencing and germ-free mouse studies, began a revolution in science. The Human Microbiome Project, run by the United States' National Institutes for Health, alongside many other studies in laboratories around the world, has revealed that we utterly depend on our microbes for health and happiness.

The human body, both inside and out, forms a landscape of habitats as diverse as any on Earth. Just as our planet's ecosystems are populated by different species of plants and animals, so the habitats of the human body host different communities of microbes. We are, like all animals, an elaborate tube. Food goes in at one end of the tube and comes out the other. We see our skin as covering our 'outside' surface, but the inner surface of our tube is also 'outside' – exposed to the environment in a similar way. Whilst the layers of our skin protect us from the elements, invading microbes and harmful substances, the cells of the digestive tract that runs right through us must also keep us safe. Our true 'inside' is not this digestive tract, but the tissues and organs, muscles and bones that are packed away between the inside and the outside of our tubular selves.

The surface of a human being, then, is not just their skin, but the twists and turns, furrows and folds of their inner tube as well. Even the lungs, the vagina and the urinary tract count as being on the outside – as part of the surface – when you view the body in this way. No matter if it's inner or outer, *all* of this surface is potential real estate for microbes. Sites vary in their value, with dense city-like communities building up in resource-rich prime locations like the intestines, while sparser collections of species occupy more 'rural' or hostile locations such as the lungs and the stomach. The Human Microbiome Project set out to characterise these communities, by sampling microbes from eighteen sites across the inner and outer surfaces of the human body, in each of hundreds of volunteers.

Over the first five years of the HMP, molecular microbiologists oversaw a biotechnological echo of the golden age of species discovery; one that saw formaldehyde-infused collecting cabinets stuffed with the haul of birds and mammals discovered and named by explorer–biologists in the eighteenth and nineteenth centuries. The human body is, as it turns out, a treasure trove of strains and species new to science, many of them present on only one or two of the volunteers participating in the study. Far from each person harbouring a replicate set of microbes, very few strains of bacteria are common to everyone. Each of us contains communities of microbes as unique as our fingerprints.

Though the finer details of our inhabitants are specific to each of us, we all play host to similar microbes at the highest hierarchical levels. The bacteria living in your gut, for example, are more similar to the bacteria in the gut of the person sitting next to you, than they are to the bacteria on your knuckles. What's more, despite our distinctive communities, the functions they perform are usually indistinguishable. What bacterium A does for you, bacterium B might do for your best friend.

From the arid, cool plains of the skin on the forearms to the warm, humid forests of the groin and the acidic, low-oxygen environment of the stomach, each part of the body offers a home to those microbes that can evolve to exploit it. Even within a habitat, distinct niches host different collections of species. The skin, all two square metres of it, contains as many ecosystems as the landscapes of the Americas, but in miniature. The occupants of the sebum-rich skin of the face and back are as different from those of the dry, exposed elbows as the tropical forests of Panama are from the rocks of the Grand Canyon. Where the face and back are dominated by species belonging to the genus *Propionibacterium*, which feed off the fats released by the densely packed pores in these areas, the elbows and forearms host a far more diverse community. Moist areas, including the navel, the underarms and the groin, are home to *Corynebacterium* and *Staphylococcus* species, which love the high humidity, and feed off the nitrogen in the sweat.

This microbial second skin provides a double layer of protection to the body's true interior, reinforcing the sanctity of the barrier formed by the skin cells. Invading bacteria with malicious intentions struggle to get a foothold in these closely guarded bodily border towns, and face an onslaught of chemical weapons when they try. Perhaps even more vulnerable to invasion are the soft tissues of the mouth, which must resist colonisation by a flood of intruders smuggled on food and floating in the air.

From the mouths of their volunteers, researchers working on the Human Microbiome Project took not just one sample, but nine, each from a slightly different location. These nine sites turned out to have discernibly different communities, within mere centimetres of one another, made up of around 800 species of bacteria dominated by *Streptococcus* species and a handful of other groups. *Streptococcus* gets bad press, on account of the many species that cause diseases, from 'strep throat' to the flesh-eating infection necrotising fasciitis. But many other species in this genus behave themselves impeccably, crowding out nasty challengers in this vulnerable entrance to the body. Of course these tiny distances between sampling locations within the mouth may seem insignificant to us, but to microbes they are like vast plains and mountain ranges with climates as different as northern Scotland and the south of France.

Imagine, then, the climatic leap from the mouth to the nostrils. The viscous pool of saliva on a rugged bedrock replaced by a hairy forest of mucus and dust. The nostrils, as you might expect from their gatekeeper status at the entrance to the lungs, harbour a great range of bacterial groups, numbering around 900 species, including large colonies of *Propionibacterium*, *Corynebacterium*, *Staphylococcus* and *Moraxella*.

Heading down the throat towards the stomach sees the enormous diversity of species found in the mouth drop dramatically. The highly acidic stomach kills many of the microbes that enter with food, and just one species is known for certain to reside there permanently in some people – *Helicobacter pylori*, whose presence may be both a blessing and a curse. From this point on, the journey through the digestive tract reveals an ever-greater density – and diversity – of microbes. The stomach opens into the small intestine, where food is rapidly digested by our very own enzymes and absorbed into the bloodstream. There are still microbes here though; around ten thousand individuals in every millilitre of gut contents at the start of this 7-m-long tube, rising to an incredible ten million per millilitre at the end, where the small intestine meets the large intestine's starting point.

Just outside the safe-house created by the appendix is a teeming metropolis of microbes, in the heart of the microbial landscape of the human body – the tennis-ball-like caecum, to which the appendix is attached. This is the epicentre of microbial life, where trillions of individual microbes of at least 4,000 species make the most of the partially digested food that has passed through round one of the nutrient-extraction process in the small intestine. The tough bits – plant fibres – are left over for the microbes to tackle in round two.

The colon, which forms most of the length of the large intestine, running up the right-hand side of your torso, across your body under your rib cage, and back down the left-hand side, provides homes for

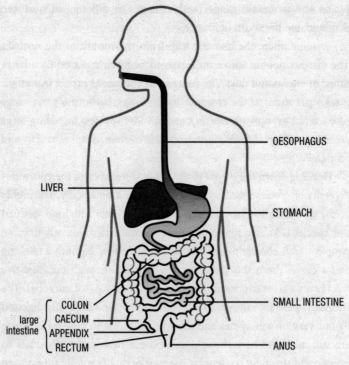

The human gut.

microbes, numbering one trillion (1,000,000,000,000) individuals per millilitre by now, in the folds and pits of its walls. Here, they pick up the scraps of our food and convert them into energy, leaving their waste products to be absorbed into the cells of the colon's walls. Without the gut's microbes, these colonic cells would wither and die – whilst most of the body's cells are fed by sugar transported in the blood, the colonic cells' main energy source is the waste products of the microbiota. The colon's moist, warm, swamp-like environment, in parts completely devoid of oxygen, provides not only a source of incoming food for its inhabitants, but a nutrient-rich mucus layer, which can sustain the microbes in times of famine.

Because HMP researchers would have to cut open their volunteers to sample the different habitats of the gut, a far more practical way of collecting information about the gut's inhabitants was to sequence the DNA of microbes found in the stool. On its passage through the gut, the food we eat is mostly digested and absorbed, both by us and our microbes, leaving only a small amount to come out the other end. Stool, far from being the remains of our food, is mostly bacteria, some dead, some alive. Around 75 per cent of the wet weight of faeces is bacteria; plant fibre makes up about 17 per cent.

At any one time, your gut contains about 1.5 kg of bacteria – that's about the same weight as the liver – and the lifespans of individuals are a matter of just days or weeks. The 4,000 species of bacteria found in the stool tell us more about the human body than all the other sites put together. These bacteria become a signature of our health and dietary status, not only as a species, but as a society, and personally. By far the most common group of bacteria in the stool are the *Bacteroides*, but because our gut bacteria eat what we eat, bacterial communities in the gut vary from person to person.

The gut microbes aren't just scavengers, though, taking advantage of our leftovers. We have exploited them too, especially when it comes to outsourcing functions that would take us time to evolve for ourselves. After all, why bother having a gene for a protein that makes Vitamin B12, which is essential for our brain function, when *Klebsiella* will do it for you? And who needs genes to shape the intestine's walls, when *Bacteroides* have them? It's much cheaper and easier than evolving them afresh. But, as we will discover, the role of the microbes living in the gut goes far beyond synthesising a few vitamins.

The Human Microbiome Project began by looking only at the microbiotas of healthy people. With this benchmark set down, the HMP went on to ask how they differ in poor health, whether our modern illnesses could be a consequence of those differences, and if so, what was causing the damage? Could skin conditions like acne, psoriasis and dermatitis signal disruption to the skin's normal balance of microbes? Might inflammatory bowel disease, cancers of the digestive tract and even obesity be due to shifts in the communities of microbes living in the gut? And, most extraordinarily, could conditions that were apparently far removed from microbial epicentres, such as allergies, autoimmune diseases and even mental health conditions, be brought on by a damaged microbiota?

Lee Rowen's educated guess in the sweepstake at Cold Spring Harbor hinted at a much deeper discovery. We are not alone, and our microbial passengers have played a far greater role in our humanity than we ever expected. As Professor Jeffrey Gordon puts it:

> This perception of the microbial side of ourselves is giving us a new view of our individuality. A new sense of our connection to the microbial world. A sense of the legacy of our personal interactions with our family and environment early in life. It's causing us to pause and consider that there might be another dimension to our human evolution.

We have come to depend on our microbes, and without them, we would be a mere fraction of our true selves. So what does it mean to be just 10 per cent human?

ONE

Twenty-First-Century Sickness

In September 1978, Janet Parker became the last person on Earth to die of smallpox. Just 70 miles from the place where Edward Jenner had first vaccinated a young boy against the disease with cowpox pus from a milkmaid, 180 years earlier, Parker's body played host to the virus in its final outing in human flesh. Her job as a medical photographer at the University of Birmingham in the UK would not have put her in direct jeopardy were it not for the proximity of her dark room to the laboratory beneath. As she sat ordering photographic equipment over the telephone one afternoon that August, smallpox viruses travelled up the air ducts from the Medical School's 'pox' room on the floor below, and brought on her fatal infection.

The World Health Organisation (WHO) had spent a decade vaccinating against smallpox around the world, and that summer they were on the brink of announcing its complete eradication. It had been nearly a year since the final naturally occurring case of the disease had been recorded. A young hospital cook had recovered from a mild form of the virus in its final stronghold of Somalia. Such a victory over disease was unprecedented. Vaccination had backed smallpox into a corner, ultimately leaving it with no vulnerable humans to infect, and nowhere to go.

But the virus did have one tiny pocket to retreat to – the Petri dishes filled with human cells that researchers used to grow and study the disease. The Medical School of Birmingham University was one such viral sanctuary, where one Professor Henry Bedson and his team were hoping to develop the means to quickly identify any pox viruses that might emerge from animal populations now that smallpox was gone from humans. It was a noble aim, and they had the blessing of the WHO, despite inspectors' concerns about the pox room's safety protocols. With just a few months left before Birmingham's lab was due to close anyway, the inspectors' worries did not justify an early closure, or an expensive refit of the facilities.

Janet Parker's illness, at first dismissed as a mild bug, caught the attention of infectious disease doctors a fortnight after it had begun. By now she was covered in pustules, and the possible diagnosis turned to smallpox. Parker was moved into isolation, and samples of fluid were extracted for analysis. In an irony not lost on Professor Bedson, his team's expertise in identifying pox viruses was called upon for verification of the diagnosis. Bedson's fears were confirmed, and Parker was moved to a specialist isolation hospital nearby. Two weeks later on 6 September, with Parker still critically ill in hospital, Professor Bedson was found dead at his home by his wife, having slit his own throat. On 11 September 1978, Janet Parker died of her disease.

Janet Parker's fate was that of many hundreds of millions before her. She had been infected by a strain of smallpox known as 'Abid', named after a three-year-old Pakistani boy who had succumbed to the disease eight years previously, shortly after the WHO's intensive smallpox eradication campaign had got under way in Pakistan. Smallpox had become a significant killer across most of the world by the sixteenth century, in large part due to the tendency of Europeans to explore and colonise other regions of the world. In the eighteenth century, as human populations grew and became increasingly mobile, smallpox spread to become one of the major causes of death around the world, killing as many as 400,000 Europeans each year, including roughly one in ten infants. With the uptake of variolation – a crude and risky predecessor of vaccination, involving intentional infection of the healthy with the smallpox fluids of sufferers – the death toll was reduced in the latter half of the eighteenth century. Jenner's discovery of vaccination using cowpox in 1796 brought further relief. By the 1950s, smallpox had been all but eliminated from industrialised countries, but there were still 50 million cases annually worldwide resulting in over 2 million deaths each year.

Though smallpox had released its grip on countries in the industrialised world, the tyrannical reign of many other microbes continued in the opening decade of the twentieth century. Infectious disease was by far the dominant form of illness, its spread aided by our human habits of socialising and exploring. The exponentially rising human population, and with that, ever-greater population densities, only eased the person-to-person leap that microbes needed to make in order to continue their life cycle. In the United States, the top three causes of death in 1900 were not heart disease, cancer and stroke, as they are today, but infectious diseases, caused by microbes passed between people. Between them, pneumonia, tuberculosis and infectious diarrhoea ended the lives of one-third of people.

Once regarded as 'the captain of the men of death', pneumonia begins as a cough. It creeps down into the lungs, stifling breathing and bringing on a fever. More a description of symptoms than a disease with a sole cause, pneumonia owes its existence to the full spectrum of microbes, from tiny viruses, through bacteria and fungi, to protozoan ('earliest-animal') parasites. Infectious diarrhoea, too, can be blamed on each variety of microbe. Its incarnations include the 'blue death' – cholera – which is caused by a bacterium; the 'bloody flux' – dysentery – which is usually thanks to parasitic amoebae; and 'beaver fever' – giardiasis, again from a parasite. The third great killer, tuberculosis, affects the lungs like pneumonia, but its source is more specific: an infection by a small selection of bacteria belonging to the genus *Mycobacterium*.

A whole host of other infectious diseases have also left their mark, both literally and figuratively, on our species: polio, typhoid, measles, syphilis, diphtheria, scarlet fever, whooping cough and various forms of flu, among many others. Polio, caused by a virus that can infect the central nervous system and destroy nerves controlling movements, paralysed hundreds of thousands of children each year in industrialised countries at the beginning of the twentieth century. Syphilis – the sexually transmitted bacterial disease – is said to have affected 15 per cent of the population of Europe at some point in their lifetime. Measles killed around a million people a year. Diphtheria – who remembers this heart-breaker? – used to kill 15,000 children each year in the United States alone. The flu killed between five and ten times as many people in the two years following the First World War than were killed fighting in the war itself.

Not surprisingly these scourges had a major influence on human life expectancy. Back then, in 1900, the average life expectancy across the whole planet was just thirty-one years. Living in a developed country improved the outlook, but only to just shy of fifty years. For most of our evolutionary history, we humans have managed to live to only twenty or thirty years old, though the *average* life expectancy would have been much lower. In one single century, and in no small part because of developments in one single decade – the antibiotic revolution of the 1940s – our average time on Earth was doubled. In 2005, the average human could expect to live to sixty-six, with those in the richest countries reaching, again on average, the grand old age of eighty.

These figures are highly influenced by the chances of surviving infancy. In 1900, when up to three in ten children died before the age of five, average life expectancy was dramatically lower. If, at the turn of the next century, rates of infant mortality had remained at the level they were in 1900, over half a million children would have died before their first birthday in the United States each year. Instead, around 28,000 did. Getting the vast majority of children through their first five years unscathed allows most of them to go on and live to 'old age' and brings the average life expectancy up accordingly.

Though the effects are far from fully felt in much of the developing world, we have, as a species, gone a long way towards conquering our oldest and greatest enemy: the pathogen. Pathogens – disease-causing microbes – thrive in the unsanitary conditions created by humans living en masse. The more of us we cram onto our planet, the easier it becomes for pathogens to make a living. By migrating, we give them access to yet more humans, and in turn, more opportunity to breed, mutate and evolve. Many of the infectious diseases we have contended with in the last few centuries originated in the period after early humans had left Africa and set up home across the rest of the world. Pathogens' world domination mirrored our own; few species have as loyal a pathogenic following as us.

For many of us living in more developed countries, the reign of infectious diseases is confined to the past. Just about all that remain of thousands of years of mortal combat with microbes are memories of the sharp prick of our childhood immunisations followed by the 'reward' of a polio-vaccine-infused sugar lump, and perhaps more clearly, the melodramatic queues outside the dinner hall as we waited with our school friends for a teenage booster shot. For many children and teenagers growing up now, the burden of history is even lighter, as not only the diseases themselves, but once-routine vaccinations, such as the dreaded 'BCG' for tuberculosis, are no longer necessary.

Medical innovations and public health measures – largely those of the late nineteenth and early twentieth centuries – have made a profound difference to life as a human. Four developments in particular have taken us from a two-generation society to a four-, or even five-generation society in just one, long, lifetime. The first and earliest of these, courtesy of Edward Jenner and a cow named Blossom, is, of course, vaccination. Jenner knew that milkmaids were protected from developing smallpox by virtue of having been infected by the much milder cowpox. He thought it possible that the pus from a milkmaid's pustules might, if injected into another person, transfer that protection. His first guinea pig was an eight-year-old boy named James Phipps – the son of Jenner's gardener. Having inoculated Phipps, Jenner went on to attempt to infect the brave lad, twice injecting pus from a true smallpox infection. The young boy was utterly immune.

Beginning with smallpox in 1796, and progressing to rabies,

typhoid, cholera and plague in the nineteenth century, and dozens of other infectious diseases since 1900, vaccination has not only protected millions from suffering and death, but has even led to countrywide elimination or complete global eradication of some pathogens. Thanks to vaccination, we no longer have to rely solely on our immune systems' experiences of full-blown disease to defend us against pathogens. Instead of acquiring natural defences against diseases, we have circumvented this process using our intellect to provide the immune system with forewarning of what it might encounter.

Without vaccination, the invasion of a new pathogen prompts sickness and possibly death. The immune system, as well as tackling the invading microbe, produces molecules called antibodies. If the person survives, these antibodies form a specialist team of spies that patrol the body looking out specifically for that microbe. They linger long after the disease has been conquered, primed to let the immune system know the moment there is a reinvasion of the same pathogen. The next time it is encountered, the immune system is ready, and the disease can be prevented from taking hold.

Vaccination mimics this natural process, teaching the immune system to recognise a particular pathogen. Instead of suffering the disease to achieve immunity, now we suffer only the injection, or oral administration, of a killed, weakened or partial version of the pathogen. We are spared illness but our immune systems still respond to the introduction of the vaccine, and produce antibodies that help the body to resist disease if the same pathogen invades for real.

Society-wide vaccination programmes are designed to bring about 'herd immunity' by vaccinating a large enough proportion of the population that contagious diseases cannot continue their spread. They have meant that many infectious diseases are almost completely eliminated in developed countries, and one, smallpox, has been totally eradicated. Smallpox eradication, as well as dropping the incidence of the disease from 50 million cases a year worldwide to absolutely none in little more than a decade, has saved governments billions in both the direct cost of vaccination and medical care, and the indirect societal costs of illness. The United States, which contributed a disproportionately large amount of money to the global eradication effort, recoups its investment every twenty-six days in unspent costs. Governmental vaccination schemes for a dozen or so other infectious diseases have dramatically reduced the number of cases, reducing suffering and saving lives and money.

Today, most countries in the developed world run vaccination programmes against ten or so infectious diseases, and half a dozen are marked for regionwide elimination or global eradication by the World Health Organisation. These programmes have had a dramatic effect on the incidence of these diseases. Before the worldwide eradication programme for polio began in 1988, the virus affected 350,000 people a year. In 2012, the disease was confined to just 223 cases in only three countries. In just twenty-five years, around half a million deaths have been prevented and 10 million children who would have been paralysed are free to walk and run. Likewise for measles and rubella: in a single decade, vaccination of these once-common diseases has prevented 10 million deaths worldwide. In the United States, as in most of the developed world, the incidence of nine major childhood diseases has been reduced by 99 per cent by vaccination. In developed countries, for every 1,000 babies born alive in 1950, around forty would die before their first birthday. By 2005, that figure had been reduced by an order of magnitude, to about four. Vaccination is so successful that only the oldest members of Western society can remember the horrendous fear and pain of these deadly diseases. Now, we are free.

After the development of the earliest vaccines came a second major health innovation: hygienic medical practice. Hospital hygiene is something we are still under pressure to improve today, but in comparison with the standards of the late nineteenth century, modern hospitals are temples of cleanliness. Imagine, instead, wards crammed full with the sick and dying, wounds left open and rotting, and doctors' coats covered in the blood and gore of years of surgeries. There was little point in cleaning – infections were thought to be the result of 'bad air', or *miasma*, not germs. This toxic mist was thought to rise from decomposing matter or filthy water – an intangible force beyond the control of doctors and nurses. Microbes had been discovered 150 years previously, but the connection had not been made between them and disease. It was believed that miasma could not be transferred by physical contact, so infections were spread by the very people charged with curing them. Hospitals were a new invention, born of a drive towards public health care and a desire to bring 'modern' medicine to the masses. Despite the good intentions, they were filthy incubators for disease, and those attending them risked their lives for the treatment they needed.

Women suffered most as a result of the proliferation of hospitals, as the risks of labour and giving birth, rather than falling, actually rose. By the 1840s, up to 32 per cent of women giving birth in hospital would subsequently die. Doctors – all male at that time – blamed their deaths on anything from emotional trauma to uncleanliness of the bowel. The true cause of this horrifyingly high death rate would at last be unravelled by a young Hungarian obstetrician by the name of Ignaz Semmelweis.

At the hospital where Semmelweis worked, the Vienna General, women in labour were admitted on alternate days into two different clinics. One was run by doctors, and the other by midwives. Every second day, as Semmelweis walked to work, he'd see women giving birth on the street outside the hospital doors. On those days, it was the turn of the clinic run by doctors to admit labouring women. But the women knew the odds for their survival would not be good if they could not hold on until the following day. Childbed fever – the cause of most of the deaths – lurked in the doctors' clinic. So they waited, cold and in pain, in the hope that their baby would delay its entrance to the world until after midnight had struck.

Getting admitted to the midwife-run clinic was, relatively speaking, a far safer proposition. Between 2 and 8 per cent of new mothers would die of childbed fever in the care of midwives – far fewer than succumbed in the doctors' clinic.

Despite his junior status, Semmelweis began to look for differences between the two clinics that might explain the death rates. He thought overcrowding and the climate of the ward might be to blame, but found no evidence of any difference. Then, in 1847, a close friend and fellow doctor, Jakob Kolletschka, died after being accidentally cut by a student's scalpel during an autopsy. The cause of death: childbed fever.

After Kolletschka's death, Semmelweis had a realisation. It was the doctors who were spreading death among the women in their ward. Midwives, on the other hand, were not to blame. And he knew why. Whilst their patients laboured, the doctors would pass the time in the morgue, teaching medical students using human cadavers. Somehow, he thought, they were carrying death from the autopsy room to the maternity ward. The midwives never touched a corpse, and the patients dying on their ward were probably those whose post-natal bleeding meant a visit from the doctor.

Semmelweis had no clear idea of the form that death was taking on its passage from the morgue to the maternity ward, but he had an idea of how to stop it. To rid themselves of the stench of rotting flesh, doctors often washed with a solution of chlorinated lime. Semmelweis reasoned that if it could remove the smell, perhaps it could remove the vector of death as well. He instituted a policy that doctors must wash their hands in chlorinated lime between conducting autopsies and examining their patients. Within a month, the death rate in his clinic had dropped to match that of the midwives' clinic.

Despite the dramatic results Semmelweis achieved in Vienna and later in two hospitals in Hungary, he was ridiculed and ignored by his contemporaries. The stiffness and stench of a surgeon's scrubs were said to be a mark of his experience and expertise. 'Doctors are gentlemen, and gentlemen's hands are clean,' said one leading obstetrician at the time, all while infecting and killing dozens of women each month. The mere notion that doctors could be responsible for bringing death, not life, to their patients caused huge offence, and Semmelweis was cast out of the establishment. Women continued to risk their lives giving birth for decades, as they paid the price of the doctors' arrogance.

Twenty years later, the great Frenchman Louis Pasteur developed the germ theory of disease, which attributed infection and illness to microbes, not miasma. In 1884, Pasteur's theory was proved by the elegant experiments of the German Nobel prize-winning doctor Robert Koch. By this time, Semmelweis was long dead. He had become obsessed by childbed fever, and had gone mad with rage and desperation. He railed against the establishment, pushing his theories and accusing his contemporaries of being irresponsible murderers. He was lured by a colleague to an insane asylum, under the pretence of a visit, then forced to drink castor oil and beaten by the guards. Two weeks later, he died of a fever, probably from his infected wounds.

Nonetheless, germ theory was the breakthrough that gave Semmelweis's observations and policies a truly scientific explanation. Steadily, antiseptic hand-washing was adopted by surgeons across Europe. Hygienic practices became common after the work of the British surgeon Joseph Lister. In the 1860s, Lister read of Pasteur's work on microbes and food, and decided to experiment with chemical solutions on wounds to reduce the risk of gangrene and septicaemia. He used carbolic acid, which was known to stop wood from rotting, to wash his instruments, soak dressings and even to clean wounds during surgery. Just as Semmelweis had achieved a drop in the death rate, so too did Lister. Where 45 per cent of those he operated on had died before, Lister's pioneering use of carbolic acid slashed mortality by two-thirds, to around 15 per cent.

Closely following Semmelweis's and Lister's work on hygienic medical practice was a third public health innovation – a development that prevented millions from becoming ill in the first place. As in many developing countries today, water-borne diseases were a major health hazard in the West before the twentieth century. The sinister forces of miasma were still at work, polluting rivers, wells and pumps. In August 1854, the residents of London's Soho district began to fall ill. They developed diarrhoea, but not as you or I might know it. This was white, watery stuff, and there was no end of it. Each person could produce up to 20 litres per day, all of which was dumped in the cesspits beneath Soho's cramped houses. The disease was cholera, and it killed people in their hundreds.

Dr John Snow, a British doctor, was sceptical of the miasma theory, and had spent some years looking for an alternative explanation. From previous epidemics, he had begun to suspect that cholera was water-borne. The latest outbreak in Soho gave him the opportunity to test his theory. He interviewed Soho residents and mapped cholera cases and deaths, looking for a common source. Snow realised that the victims had all drunk from the same water pump on Broad Street (now Broadwick Street) at the heart of the outbreak. Even deaths further afield could be traced back to the Broad Street pump, as cholera was carried and passed on by those infected there. There was one anomaly: a group of monks in a Soho monastery who got their water from the same pump were completely unaffected. It was not their faith that had afforded them protection, though, but their habit of drinking the pump's water only after they had turned it into beer.

Snow had looked for patterns – connections between those who had become ill, reasons why others had escaped, links explaining the appearance of the disease outside its Broad Street epicentre. His rational study used logic and evidence to unravel the outbreak and trace its source, eliminating red herrings and accounting for anomalies. His work led to the disabling of the Broad Street pump and the subsequent discovery that a nearby cesspit had overflowed and was contaminating the water supply. It was the first-ever epidemiological study – that is, it used the distribution and patterns of a disease to understand its source. John Snow went on to use chlorine to disinfect the water supplying the Broad Street pump, and his chlorination methods were quickly put to use elsewhere. As the nineteenth century came to a close, water sanitation had become widespread.

As the twentieth century unfolded, all three public health innovations became more and more sophisticated. By the end of the Second World War, a further five diseases could be prevented through vaccination, taking the total to ten. Medical hygiene techniques were adopted internationally, and chlorination became a standard process in water-treatment plants. The fourth and final innovation to put an end to the reign of microbes in the developed world began with one world war and concluded with the second. It was the result of the hard work, and good fortune, of a handful of men. The first of these, the Scottish biologist Sir Alexander Fleming, is famously credited with 'accidentally' discovering penicillin in his laboratory at St Mary's Hospital in London. In fact, Fleming had been hunting for anti-bacterial compounds for years.

During the First World War he had treated wounded soldiers on the Western Front in France, only to see many of them die from sepsis. When the war came to an end and Fleming returned to the UK, he made it his mission to improve upon Lister's antiseptic car-bolic acid dressings. He soon discovered a natural antiseptic in nasal mucus, which he called lysozyme. But, as with carbolic acid, it could not penetrate beneath the surface of wounds, so deep infec-tions festered. Some years later, in 1928, Fleming was investigating staphylococci bacteria – responsible for boils and sore throats – when he noticed something odd on one of his Petri dishes. He had been on holiday, and had returned to a messy lab bench full of old bac-terial cultures, many of which had been contaminated with moulds. As he sorted through them, he noticed one dish in particular. Sur-rounding a patch of *Penicillium* mould was a clear ring, completely free of the staphylococci colonies that covered the remainder of the plate. Fleming spotted its significance: the mould had released a 'juice' that had killed the bacteria around it. That juice was penicillin.

Though growing the *Penicillium* had been unintentional, Fleming's recognition of its potential importance was anything but accidental. It began a process of experimentation and discovery that would span two continents and twenty years, and revolutionise medicine. In 1939, a team of scientists at Oxford University, led by the Australian pharmacologist Howard Florey, thought they could make more use of penicillin. Fleming had struggled to grow significant quantities of the mould, or to extract the penicillin it produced. Florey's team man-aged it, isolating small amounts of liquid antibiotic. By 1944, with the financial support of the War Production Board in the United States, penicillin was produced in sufficient quantities to meet the needs of soldiers returning from the D-Day invasion of Europe. Sir Alexander Fleming's dream of beating the infections of the war wounded was realised, and the following year he, Florey, and one other member of the Oxford team, Sir Ernst Boris Chain, received the Nobel Prize in Medicine or Physiology.

Over twenty varieties of antibiotics have subsequently been developed, each attacking a different bacterial weakness, and providing our immune systems with backup when they are overwhelmed by infection. Before 1944, even scratches and grazes could mean a frighteningly high chance of death by infection. In 1940, a British policeman in Oxfordshire called Albert Alexander was scratched by a rose thorn. His face became so badly infected that he had to have his eye removed, and he was on the verge of death. Howard Florey's wife Ethel, who was a doctor, persuaded Florey that Constable Alexander should become the first recipient of penicillin.

Within twenty-four hours of being injected with a tiny quantity of penicillin, the policeman's fever dropped, and he began to recover. The miracle was not to be, however. A few days into his treatment, penicillin supplies ran out. Florey had attempted to extract any remaining penicillin from the constable's urine to continue the treatment, but on the fifth day, the policeman died. It is unthinkable now to die from a scratch or an abscess, and we often take antibiotics without heed to their life-saving properties. Surgery, too, would carry enormous risk were it not for the protective shield of intravenous antibiotics given before the first cut is made.

Our twenty-first-century lives are a kind of sterile ceasefire, with infections held at bay through vaccinations, antibiotics, water sanitation and hygienic medical practice. We are no longer threatened by acute and dangerous bouts of infectious disease. Instead, the past sixty years have seen a collection of previously rare conditions rise to prominence. These chronic 'twenty-first-century illnesses' have become so common that we accept them as a normal part of being human. But what if they are not 'normal'?

Looking around among your friends and family, you won't see small-pox, measles or polio any more. You might think how lucky we are; how healthy we are these days. But look again and you might see things differently. You might see the sneezing and red, itchy eyes of your daughter's hay fever in the spring. You might think of your sister-in-law, who has to inject herself with insulin several times a day because of her type 1 diabetes. You might be worried your wife will end up in a wheelchair with multiple sclerosis as her aunt did. You might have heard about your dentist's little boy who screams, and rocks himself, and won't make eye contact, now that he has autism. You might get impatient with your mother who is too anxious to do the shopping. You might be searching for a washing powder that doesn't make your son's eczema worse. Your cousin might be the awkward one at dinner who can't eat wheat because it gives her diarrhoea. Your neighbour might have slipped unconscious whilst searching for his EpiPen after accidentally eating nuts. And *you* might have lost the battle to keep your weight where beauty magazines, and your doctor, say it should be. These conditions – allergies, autoimmune diseases, digestive troubles, mental health problems and obesity – are the new *normal*.

Let's take allergies. Perhaps there's nothing alarming about your daughter's hay fever, as 20 per cent of her friends also snuffle and sneeze their way through summer. You are not surprised by your son's eczema, because one in five of his classmates have it too. Your neighbour's anaphylactic attack, terrifying though it was, is common enough that all packaged foods carry warnings if they 'may contain nuts'. But have you ever asked yourself why one in five of your children's friends have to take an inhaler to school in case they suffer an asthma attack? Being able to breathe is fundamental to life, yet without medication, millions of children would find themselves gasping for breath. What about why one in fifteen children are allergic to at least one type of food? Can that be normal?

Allergies affect nearly half of us in developed countries. We dutifully take our antihistamines, avoid picking up the cat, and check the ingredients lists of everything we buy. We unthinkingly do what is necessary to stop our immune systems overreacting to the most ubiquitous and innocuous of substances: pollen, dust, pet hair, milk, eggs, nuts, and so on. These substances are being treated by the body as if they are germs that need to be attacked and removed. It hasn't always been this way. In the 1930s, asthma was rare, affecting perhaps one child in every school. By the 1980s, it had shot up, and one child in every *class* was affected. In the last decade or so, the rise has levelled off, but it has left a quarter of children with asthma. The same goes for other allergies: peanut allergies, for instance, trebled in just ten years at the end of the last century, and then doubled again in the next five years. Now we have nut-free zones in schools and workplaces. Eczema and hay fever too were once rare and are now a fact of life.

This is not normal.

What about autoimmune diseases? Your sister-in-law's insulin habit is common enough, with type 1 diabetes affecting, as it does, about 4 in every 1,000 people. Most people have heard of the multiple sclerosis (or MS) that's destroyed your wife's aunt's nerves. And then there's rheumatoid arthritis wrecking joints, coeliac disease attacking the gut, myositis shredding muscle fibres, lupus pulling apart cells at their core, and about eighty other such conditions. As with allergies, the immune system has gone rogue, attacking not just the germs that bring disease, but the body's own cells. You might be surprised to learn that among them, autoimmune diseases affect nearly 10 per cent of the population in the developed world.

Type 1 diabetes (T1D) makes for a great example, because it is an unmistakable condition, so records are relatively reliable. 'Type 1' is the version of diabetes that usually strikes early, often in the teenage years, attacking the cells of the pancreas, and completely preventing the production of the hormone insulin. (In type 2 diabetes, insulin is produced, but the body has grown less sensitive to it, so it doesn't work as well.) Without insulin, any glucose in the blood – whether that's from the simple sugars in sweets and desserts or from the carbohydrates in pasta and bread – cannot be converted and stored. It builds up and quickly becomes toxic, bringing with it a raging thirst and constant need to urinate for the unfortunate teenager. The patient wastes away, and weeks or months later, death follows, often from kidney failure. That is, unless insulin is injected. Pretty serious, then.

Fortunately, compared to most conditions, it's straightforward to diagnose, and always has been. These days, a quick check of the amount of glucose in the blood after fasting usually gives it away, but even 100 years ago diabetes could be detected by a willing doctor. I say willing, because the test for it involved tasting the patient's urine. A sweetness within the tang indicated that there was so much glucose in the blood that it had been forced out into the urine by the kidneys. Though undoubtedly more cases were missed in the past than now, and many would have gone unrecorded, our understanding of the prevalence of type 1 diabetes over time is a reliable indicator of the changing status of autoimmune diseases.

About 1 in 250 people in the West are stuck playing the role of their own pancreas, calculating how much insulin they need and then injecting it, to store away the glucose they have consumed. What's extraordinary is that this high prevalence is new: type 1 diabetes was almost non-existent in the nineteenth century. Hospital records for Massachusetts General Hospital in the US, kept over seventy-five years until 1898, log only twenty-one cases of diabetes diagnosed in childhood, out of nearly 500,000 patients. It's not a case of missed diagnosis, either – that urine-taste test, the rapid weight loss and the inevitable fatal outcome made the disease easy to recognise even back then.

Once formal records had been set up just before the Second World War, the prevalence of type 1 diabetes could be tracked. Around 1 or 2 children in every 5,000 were affected in the US, UK and Scandinavia. The war itself altered nothing, but not long afterwards, something changed, and cases began to rise. By 1973, diabetes was six or seven times as common as it had been in the Thirties. In the Eighties, the rise levelled off at its current figure of about 1 in 250.

The rise in diabetes is matched by rises in other autoimmune conditions. Multiple sclerosis destroyed the nervous systems of twice as many people at the turn of the millennium as it did two decades previously. Coeliac disease, in which the presence of wheat prompts the body to attack the cells of the intestine, is a startling thirty or forty times as common now as it was in the 1950s. Lupus, inflammatory bowel disease and rheumatoid arthritis too have been on the rise.

This is not normal.

What about our collective battle with excess weight? Odds are I'm right in my flippant assumption that you struggle with your weight, as well over half of us in the Western world are either overweight or obese. It's astonishing to think that being a healthy weight puts you in the minority now. Being fat is so typical that old shop mannequins have been replaced by larger versions, and television shows turn weight loss into a game. These changes are perhaps to be expected: statistically speaking, being overweight *is* the reality for most people.

But it didn't use to be. To us now, looking back at black and white photographs of the skinny young men and women of the Thirties and Forties enjoying a spell of hot weather in shorts and swimwear, these healthy people appear emaciated, with prominent ribs and lean bellies. But they are not – they are simply not carrying our modern baggage. At the start of the twentieth century, human body weights were uniform enough that few thought to keep records. But, prompted by a sudden rise in weight gain in the 1950s at the epicentre of the obesity epidemic – America – the government began keeping track. In the first national survey in the early 1960s, 13 per cent of adults were already obese. That is, they had a Body Mass Index (weight in kilograms divided by height in metres squared) of over 30. A further 30 per cent were overweight (a BMI between 25 and 30).

By 1999, the proportion of obese American adults had more than doubled to 30 per cent, and many previously healthy adults had piled on the pounds, keeping the overweight category at a plump 34 per cent. That's a total of 64 per cent overweight or obese. Trends in the UK followed the same pattern, with a bit of a lag: in 1966, just 1.5 per cent of the adult population were obese and 11 per cent were overweight. By 1999, 24 per cent were obese and 43 per cent overweight – that's 67 per cent of people now heavier than they should be. Obesity is not just about excess weight, either. It can lead to type 2 diabetes, heart disease and even some cancers, all of which are increasingly common.

You don't need me to tell you, this is not normal.

Tummy troubles too are on the rise. Your cousin may be awkward for trying out a gluten-free diet, but she's possibly not the only one at the table who suffers from irritable bowel syndrome, which affects up to 15 per cent of people. The name implies a similar level of discomfort to a midge bite, and belies the ruinous impact of the condition on the quality of life of its sufferers. Proximity to a toilet takes priority over more meaningful pursuits for most sufferers, and a near-absence of need for one makes pursuit of anything but colonic relief worthless for the remaining patients. Inflammatory bowel diseases like Crohn's disease and ulcerative colitis too are on the rise, leaving the worst affected with a bowel so damaged it has to be replaced by a colostomy bag outside the body.

This is definitely not normal.

And finally we come to mental health conditions. Your dentist's autistic son has more company than ever before, as 1 in 68 children (but 1 in every 42 boys) are on the autistic spectrum. Back in the early 1940s autism was so rare it hadn't even been given a name. Even by the time records began in 2000, it was less than half as common as it is now. You'd be right in thinking that at least some of these extra cases are due to increasing awareness and perhaps some over-diagnosis, but most experts agree that the rise in autism prevalence is genuine – something has changed. Attention deficit disorders, Tourette's syndrome and obsessive–compulsive disorder are all also on the rise. Depression and anxiety disorders too.

This increase in mental suffering is not normal.

Except these conditions *are* now so very 'normal', you might not even have realised that they are *new* illnesses, rarely encountered by our great-grandparents and those before them. Even doctors are often unaware of the histories of the conditions they treat, having received their medical training only in the context of today's doctors' experiences. As with the rise in cases of appendicitis, a change forgotten by today's medics, what matters most to front-line carers is the patients in their charge and the treatments available to them. Understanding the provenance of illness is not their responsibility, and as such, changes in prevalence are incidental to them.

In the twenty-first century, life is different thanks to the four public health innovations of the nineteenth and twentieth centuries, and so disease is different too. But our twenty-first-century illnesses are not simply another layer of ill-health, hidden beneath infectious disease, but an alternative set of conditions, created by the way we live now. At this point you might be wondering how these illnesses can possibly have something in common, such a disparate group do they seem. From the sneezing and itching of allergies, to the self-destruction of autoimmunity, the metabolic misery of obesity, the humiliation of digestive disorders and the stigma of mental health conditions, it's as if our bodies, in the absence of infectious diseases, have turned on themselves.

We could accept our new fate and be grateful that we will, at least, live long lives free from the tyranny of the pathogen. Or we could ask what has changed. Could there be a link between conditions that seem unrelated, like obesity and allergies, irritable bowel syndrome and autism? Does the shift from infectious diseases to this new set of illnesses indicate that our bodies *need* infections to stay balanced? Or is the correlation between declining infectious disease and rising chronic illness merely hinting at a deeper cause?

We are left with one big question: *Why* are these twenty-first-century illnesses happening?

At the moment, it's fashionable to look to genetics for the source of disease. The Human Genome Project has unearthed a whole heap of genes that, when mutated, can result in illness. Some mutations guarantee disease: a change to the code of the *HTT* gene on chromosome 4, for example, will always result in Huntington's disease. Others mutations simply increase the likelihood: misspellings in the genes *BRCA1* and *BRCA2* raise a woman's risk of breast cancer to up to an eight in ten chance in her lifetime, for instance.

Although this is the era of the genome, we cannot blame our DNA alone for the rise in our modern diseases. While one person might carry a version of a gene that makes them, say, more likely to become obese, that gene variant could not become dramatically more common in the population as a whole inside a single century. Human evolution just does not progress that rapidly. Not only that, but gene variants only grow more common through natural selection if they are beneficial, or their detrimental effects are suppressed. Asthma, diabetes, obesity and autism bring few advantages to their victims.

With genetics excluded as the cause of the rise, our next question must be: Has something changed in our environment? Just as a person's height is a result of not only their genes, but their environment – nutrition, exercise, lifestyle and so on – so is their disease risk. And this is where it gets complicated, as so very many aspects of our lives have changed in the last century, and pinpointing which are causes and which are mere correlations requires the patient process of scientific evaluation. For obesity and its related illnesses, changes in the way we eat are clear to see, but how this affects other twenty-first-century illnesses is less obvious.

The diseases in question offer up few clues as to their joint origin. Could the same changes in our environments that lead to obesity also generate allergies? Can there really be a common cause of mental health conditions like autism and obsessive–compulsive disorder, and gut disorders like irritable bowel syndrome?

Despite the disparities, two themes emerge. The first, clearly binding allergies and autoimmune diseases, is the immune system. We are looking for a culprit which has interfered with the immune system's ability to determine our bodily threat level, making overreactions all too common. The second theme, often hidden behind more socially acceptable symptoms, is gut dysfunction. For some modern illnesses, the link is clear: IBS and inflammatory bowel disease have bowel disturbances at the core of their presentation. For others, although it is less overt, the connection is still there. Autistic patients struggle with chronic diarrhoea; depression and IBS go hand in hand; obesity has its origin in what passes through the gut.

These two themes, the gut and the immune system, might also seem unrelated, but a closer look at the anatomy of the gut provides a further clue. Asked about their immune system, most people might think of white blood cells and lymph glands. But that's not where most of the action is. In fact, the human gut has more immune cells than the rest of the body put together. Around 60 per cent of the immune system's tissue is located around the intestines, particularly along the final section of the small intestine and into the caecum and the appendix. It's easy to think of the skin as the barrier between us and the outside world, but for every square centimetre of skin, you have two square *metres* of gut. Though it's on the 'inside', the gut has just a single layer of cells between what's essentially the outside world, and the blood. Immune surveillance along the intestines, therefore, is critical – every molecule and cell that passes by must be assessed and quarantined if necessary.

Although the threat of infectious disease has all but gone, our immune systems are still under fire. But why? Let's turn to the technique pioneered by Dr John Snow during Soho's cholera outbreak of 1854: epidemiology. Since Snow first applied logic and evidence to unravelling the mystery of the source of cholera, epidemiology has become a mainstay of medical sleuthing. It couldn't be simpler: we ask three questions: (1) *Where* are these diseases occurring? (2) *Who* are they affecting? and (3) *When* did they become a problem? The answers provide us with clues that can help us to answer the overall question: *Why* are twenty-first-century illnesses happening?

The map of cholera cases that John Snow produced in answer to *Where?* gave away cholera's likely epicentre – the Broad Street pump.

Without much detective work, it's clear to see that obesity, autism, allergies and autoimmunity all began in the Western world. Stig Bengmark, professor of surgery at University College London, puts the epicentre of obesity and its related diseases in the southern states of the US. 'States like Alabama, Louisiana and Mississippi have the highest incidence of obesity and chronic diseases in the US and the world,' he says. 'These diseases spread, with a pattern similar to a tsunami, across the world; to the west to New Zealand and Australia, to the north to Canada, to the east to Western Europe and the Arab world and to the south, particularly Brazil.'

Bengmark's observation extends to the other twenty-first-century illnesses – allergies, autoimmune diseases, mental health conditions and so on – all of which have their origin in the West. Of course geography alone does not explain the rise; it merely gives clues as to other correlates, and with luck, the cause. The clearest correlate of this particular topography of illness is wealth. A great accumulation of evidence points to the correlation between chronic diseases and affluence, from grand-scale comparisons of the gross national product of entire countries, to contrasts between socio-economic groups living in the same local area.

In 1990, the population of Germany provided an elegant natural experiment into the impact of prosperity on allergies. After four decades apart, East and West Germany were reunifying following the fall of the Berlin Wall the previous year. These two states had much in common; they shared a location, a climate, and populations composed of the same racial groups. But whilst those living in West Germany had prospered, eventually catching up and keeping pace with the economic developments of the Western world, East Germans had existed in a state of suspended animation since the Second World War and were significantly poorer than their West German neighbours. This difference in wealth was somehow related to a difference in health. A study by doctors at Munich University's Children's Hospital found that the richer West German children were twice as likely to have allergies, and three times as likely to suffer from hay fever.

This is a pattern that repeats itself for many allergic and autoimmune conditions. American children living in poverty are historically less likely to suffer from food allergies and asthma than their wealthier counterparts. The children of 'privileged' families in Germany, as judged by their parents' educations and professions, are significantly more likely to suffer from eczema than those from less privileged backgrounds. Children from impoverished homes in Northern Ireland are not as prone to developing type 1 diabetes. In Canada, inflammatory bowel disease more often accompanies a high salary than a low one. The studies go on and on, and the trends are far from local. Even a country's gross national product can be used to predict the extent of twenty-first-century illnesses within its population.

The rise of so-called Western illnesses is no longer limited to Western countries. With wealth comes chronic ill-health. As developing countries play economic catch-up, the diseases of civilisation spread. What began as a Western problem threatens to engulf the rest of the planet. Obesity tends to lead the way, and has affected large swathes of the population already, including those in developing countries. Its collective of associated conditions such as heart disease and type 2 diabetes (an insensitivity to insulin, rather than a lack of it) are trailing not far behind. Allergic disorders, including asthma and eczema, are also at the forefront of the spread, with rises under way across middle-income countries in South America, Eastern Europe and Asia. Autoimmune diseases and behavioural conditions appear to lag the most, but are now particularly common in the upper-middle-income countries, including Brazil and China. Just as many of our modern illnesses reach a plateau in the wealthiest countries, these conditions begin their ascent elsewhere.

When it comes to twenty-first-century illnesses, money is dangerous. The size of your salary, the wealth of your neighbourhood and status of your country all contribute to your risk. But of course, simply being rich does not make you ill. Money may not buy happiness, but it does buy clean water, freedom from infectious disease, calorie-rich foods, an education, a job in an office, a small family, holidays to far-flung places, and many other luxuries besides. Asking *Where?* tells us not just the location of our modern plagues, but that it is money that's bringing us chronic ill-health.

Intriguingly though, this relationship between increasing wealth and poorer health disconnects at the very richest end of the scale. The wealthiest people in the wealthiest countries appear to be better able to lift themselves clear of the chronic disease epidemic. What begins as a preserve of the rich (think tobacco, takeaway food and ready meals) ends as the staple of the poor. Meanwhile, the well-off gain access to the latest health information, the best health care, and the freedom to make choices that keep them well. Now, while the richest cohorts of society in developing countries gain weight and acquire allergies, it is the *poorest* in developed countries who are more and more likely to be overweight and to suffer from chronic ill-health.

Next, we must ask *Who?* Does wealth and a Western lifestyle bring ill-health to everyone, or are some groups affected more than others? It's a pertinent question: in 1918, as many as 100 million people died from the flu pandemic that swept the globe after the First World War. Asking *Who?* provided an answer, that, with today's medical knowledge, could potentially have considerably reduced the death toll. Whereas flu usually kills vulnerable members of society – the young, the old, and the already sick – the 1918 flu killed mainly healthy young adults. These victims, in the prime of their lives, are likely to have died not from the flu virus itself, but from the 'cytokine storm' unleashed by their immune systems in an attempt to clear the virus. The cytokines – immune messenger chemicals which ramp up the immune response – can inadvertently lead to a reaction that's more dangerous than the infection itself. The younger and fitter the patients, the greater the storm their immune systems created, and the more likely they were to die from the flu. Asking *Who?* tells us something of what made this particular flu virus so dangerous, and would have enabled us to direct medical care not just at fighting the virus, but also towards calming the storm.

Who? is composed of three elements. What *age* are those affected by twenty-first-century illnesses? Are there differences in how these conditions affect people of different *races*? And are the *sexes* affected equally?

Let's start with age. It's easy to assume that diseases associated with developed, wealthy countries, where health care is good, are an inevitable consequence of our ageing population. *Of course new diseases are on the rise!* you might think. *We live so long now!* Surely so many of us living well into our seventies and eighties guarantees a whole host of new health challenges? Of course, as we release ourselves from the burden of death-by-pathogen, we will inevitably suffer from death-by-something-else, but many of the illnesses we face now are not simply diseases of old age, released by our longer life expectancy. Unlike cancer, whose rise is at least partly attributable to the cellular replacement process breaking down in older bodies, twenty-first-century illnesses are not all old-age-related. In fact, most of them show a preference for children and young adults, despite being relatively rare among these age groups during the age of infectious disease.

Food allergies, eczema, asthma and skin allergies often begin at birth or in the first few years of a child's life. Autism typically presents itself in toddlers, and is diagnosed before the age of five. Autoimmune diseases can hit at any time, but many show themselves at a young age. Type 1 diabetes, for example, typically reveals itself in childhood and the early teens, though it can also crop up in adulthood. Multiple sclerosis, the skin condition psoriasis, and inflammatory bowel diseases such as Crohn's disease and ulcerative colitis, all typically attack in the twenties. And lupus usually affects people between the ages of fifteen and forty-five. Obesity too, is a disease that can start young, with around 7 per cent of American babies considered over the normal weight at birth, rising to 10 per cent by the time they are toddlers, and about 30 per cent becoming overweight later in childhood. Older people are not immune to twenty-first-century illnesses – almost all of them can strike suddenly at any age – but the fact that they so often affect the young suggests it is not the ageing process itself that triggers them.

Even among those diseases that kill people in the West in 'old' age

– heart attacks, strokes, diabetes, high blood pressure and cancers – most have their roots in weight gain that begins in childhood or early adulthood. We can't attribute deaths from these conditions to our longer lifespans alone, as even those people in traditional societies who make it to eighty or ninety years old very rarely die of this set of 'age-related' illnesses. Twenty-first-century illnesses are not limited by the burgeoning top tier of our demographic ranks, but rather are hitting us, like the 1918 flu, in what should be the prime of our lives.

On to race. The Western world – North America, Europe and Australasia – is a largely white place, so are our new health problems actually a genetic predisposition among white people? In fact, within these continents, whites do not consistently have the highest rates of obesity, allergies, autoimmunity or autism. Blacks, Hispanics and South Asians tend to have higher incidences of obesity than whites, and allergies and asthma disproportionately affect blacks in some areas and whites in others. No clear pattern emerges for autoimmune diseases, with some, such as lupus and scleroderma, affecting blacks more, and others, including childhood diabetes and multiple sclerosis, tending to prefer whites. Autism does not appear to affect races differently, though black children are often diagnosed later.

Could what seem like racial differences actually be largely due to other factors, such as wealth or location, rather than to the genetic tendencies of each race? In an elegantly designed statistical study, the higher rate of asthma in black American children than in other races was found to be due not to race itself, but the greater tendency of black families to live in inner-city urban locations, where asthma is more common in all children. Rates of asthma among black children growing up in Africa are, as in most less developed regions, low.

A neat way to untangle the effects of ethnicity and environment in bringing on twenty-first-century illnesses is to look at the health of migrants. In the 1990s, civil war led to a large exodus of families from Somalia to Europe and North America. Having escaped turmoil in their own country, the Somali diaspora faced a fresh battle. Whereas rates of autism are extremely low in Somalia, the incidence in children born to Somali migrants rapidly jumped to match that of non-migrant children. Among the large Somali community of Toronto, Canada, autism is referred to as 'the Western disease', as so many migrant families are affected by it. In Sweden too, children of immigrants from Somalia have three or four times the rate of autism as Swedish children. Race, then, seems far less important than location.

So what about the final aspect of *Who?*: sex. Do women and men suffer equally? That women have stronger immune systems may not come as a surprise to anyone who has witnessed a bout of 'man flu'. But unfortunately, in this immune-mediated epidemic of chronic ill-health, women's immune superiority proves a disadvantage. While men seem to succumb to the most benign of colds, women battle demons that only their immune systems can see.

Autoimmune diseases show the widest divergence, with the vast majority of disorders affecting more women than men, though several affect both sexes equally, and a couple show a preference for males. Allergies, though more common in boys than girls, affect more women than men after puberty. Gut disorders too affect more women than men – just slightly so for inflammatory bowel disease, but twice as many women have irritable bowel syndrome.

Perhaps surprisingly, obesity also seems to affect women more than men, particularly in developing countries. But when measurements other than BMI are used, such as waist circumference, these suggest that men and women actually suffer equally from dangerous levels of excess weight. Likewise, although it appears that some mental health conditions, including depression, anxiety and obsessive–compulsive disorder, affect more women than men, part of this difference may be down to the male reluctance to admit to feeling blue. In autism, it is males who carry the burden, with five times as many boys affected as girls. Perhaps in autism, as with allergies, which tend to strike young, and those autoimmune diseases that begin in childhood, the pre-pubertal onset makes all the difference. Without the influence of adult sex hormones, these illnesses are not subject to the same female bias.

Women's strong immune systems are likely to be behind the female preponderance of several twenty-first-century illnesses. For conditions that involve overreactions of the immune system, such as allergies and autoimmunity, a stronger starting point is likely to lead to a greater response. Sex hormones, genetics and lifestyle differences could all play a role too – the jury's out on exactly why women are worse affected. Whatever it is, the female bias in these modern plagues emphasises the immune system's underlying role in their development. Twenty-first-century illnesses are not diseases of old age. They are not diseases of genetic inheritance. They are diseases of the young, the privileged, and those of immune fortitude, especially women.

We have reached the final question of our epidemiological mystery: *When?* Arguably, this is the most important question of all. I have been calling the modern chronic disease epidemic one of *twenty-first-century* sickness, though its root is not in this young century, but the last. What a century it was, the twentieth, bringing some of the greatest innovations and discoveries of all of human history. But over the course of its one hundred years, following the near-elimination of serious infectious disease in the developed world, came a new set of illnesses which went from being exceptionally rare to remarkably common. Among the many developments that took place in the last century lies the change, or cluster of changes, that have caused this rise. Pinpointing the moment that the rise began could provide our greatest clue as to its origin.

You may have got a feel for the timings already. In the US, a sharp upturn in type 1 diabetes cases began around the mid-century. Analysis of conscript data in both Denmark and Switzerland placed it in the early 1950s, in the Netherlands in the late 1950s, and in slightly less developed Sardinia in the 1960s. Rises in asthma and eczema started in the late Forties and early Fifties, and increases in Crohn's disease and multiple sclerosis took off in the Fifties. Trends in obesity were first recorded on a large scale in the Sixties, making it difficult to determine the start of the epidemic as we see it now, but some experts point to the end of the Second World War in 1945 as a likely turning point. A sharp upturn in cases of obesity took off in the 1980s, but the origin of the rise certainly occurred before then. Similarly, the number of children diagnosed with autism each year was not recorded until the late Nineties, but the condition was first described in the mid-1940s.

Something changed around the middle of the last century. Perhaps more than one thing changed, and perhaps it continued to change in the decades that followed. That change has spread around the world since, enveloping ever more countries as the decades go by. To find the cause of our twenty-first-century illnesses, we must look at the changes centred on one extraordinary decade: the 1940s.

From asking *What?*, *Where?*, *Who?* and *When?*, we have established four things. First, our twenty-first-century illnesses often arise in the gut, and are associated with the immune system. Second, they strike young, often in children, teenagers and young adults, and many affect more women than men. Third, these illnesses occur in the Western world, but are now on the rise in developing countries as they modernise. Fourth, the rise began in the West in the 1940s, and developing countries followed suit later.

And so we return to the big question: *Why* have these twenty-first-century illnesses taken over? What is it about our modern, Western, wealthy lives that is making us so chronically ill?

As individuals and as a society, we have gone from frugal to indulgent; from traditional to progressive; from lacking luxuries to being bombarded by them; from poor health care to excellent medical services; from a budding to a blooming pharmaceutical industry; from active to sedentary; from provincial to globalised; from make-do-and-mend to refresh-and-replace; and from prudish to uninhibited.

Amongst these changes, and in answer to our mystery, are 100 trillion tiny clues waiting to be followed.

TWO

All Diseases Begin in the Gut

Garden warblers are the very epitome of the birder's greatest identification challenge – the LBJ, or 'Little Brown Job'. Their most distinguishing feature is, in fact, the absolute lack of any distinguishing features, making recognition of this small bird through a pair of shakily held binoculars particularly difficult. But boring birds these are not. Just a few months after hatching, juvenile garden warblers embark on a 4,000-mile migration from their summer homes across Europe to their winter residences in sub-Saharan Africa. It is a route they have never taken before, and they do it without the help of either their more experienced parents, or a map.

Before these tiny birds head off on this incredible journey, they prepare themselves for the effort of flying and the lack of food en route by becoming fat. Over just a couple of weeks, the warblers double in weight, going from a slender 17 g to a distinctly portly 37 g. In human terms, they become morbidly obese. On each day of the pre-migration binge, a garden warbler gains around 10 per cent of its original body weight – the equivalent of a 10-stone man putting on *a stone a day* until he weighs 22 stone. Then, once the birds are plump enough, they undertake a feat of endurance beyond the imagination of most elite athletes – flying thousands of miles with just a handful of meals along the way.

Of course, to become that fat that fast, the warblers must gorge themselves on summer's bounty of food. Practically overnight, the birds shift from a diet of insects to one of berries and figs. Although the fruit is ripe enough to eat for several weeks before their binge begins, the warblers leave it untouched until the time is right. It's as if a switch flips inside them, and suddenly they dedicate themselves to eating.

For a long time, researchers assumed that the weight gain in warblers and other migratory birds was simply the consequence of *hyperphagia* – excessive eating. But the incredible speed of the shift in these birds from lean to morbidly obese suggested there was something else going on to help them store so much fat. Something that had less to do with how much food they ate, and more to do with how that food was stored in their bodies. By keeping a check on how many extra calories the warblers ate, and how many calories came out in their droppings, researchers realised that the additional food the birds were consuming did not fully account for the weight they were managing to gain.

The riddle continues when it comes to the birds losing the excess weight again. Of course, as the obese warblers make their journey across the Mediterranean Sea and the Sahara Desert, their fat supplies dwindle. By the time they have arrived and settled in to their African winter home, they have returned to a normal warbler weight. But here's the strange thing: *captive* garden warblers are no different. During the pre-migratory period at the end of summer, these caged birds still gain weight, becoming thoroughly obese in preparation for a journey they will never make. And, at the exact point that wild warblers arrive at their destination, the captive warblers completely shed their excess fat. Despite *not* flying 4,000 miles, and having unlimited access to food, these captive birds still lose the weight again when the migratory period is over.

It is quite extraordinary that warblers deprived of cues in the weather, day length and seasonal food supplies are still able to rapidly gain enormous stores of fat for the migratory period and then slim down again apparently effortlessly, perfectly in sync with their wild cousins. These are birds, with brains the size of a pea. They don't gain weight, then think to themselves: 'I really must go on a diet.' They don't fast, or exercise madly, either. Their food intake does relapse after the binge, but again, not enough to account for losing that much weight, that fast. Imagine being able to drop a stone a day for seven days – that's the degree of weight loss these little birds manage once the migratory period is over. Even eating nothing at all would not result in that kind of weight loss in humans.

Although we don't yet know exactly how this astounding degree of weight change is regulated in the warblers' bodies, the fact that these shifts happen beyond what is expected from changes in caloric intake makes one thing clear: maintaining a stable weight is not always a simple case of balancing calories-in and calories-out. In humans, the scientifically accepted explanation for weight gain is this: 'The fundamental cause of obesity and overweight is an energy imbalance between calories consumed and calories expended.'

It seems obvious: if you eat too much and move too little, the extra energy must be stored and you will gain weight. And if you want to lose weight, you must eat less and move more. But the warblers are able to rapidly lay down fat reserves that appear to go far beyond the calories they eat, and then deplete those reserves far beyond the calories they burn. Clearly, there's more to the weight-regulation game than meets the eye. If calories-in versus calories-out isn't true for warblers, perhaps it's not true for humans either?

Attempting to treat over 10,000 cases of obesity made the Indian physician Dr Nikhil Dhurandhar wonder the exact same thing. His patients returned again and again, after regaining the small amounts of weight they'd lost, or failing to lose any weight whatsoever. Despite the difficulties, Dhurandhar and his father – another doctor specialising in obesity – ran one of the most successful obesity clinics in Mumbai in the 1980s. But after a decade of trying to help people to eat less and move more, he began to feel his efforts – and those of his patients – were futile. 'After weight loss, you gain weight again: that is the big problem. And that has been my frustration.' Dhurandhar wanted to know more about mechanisms behind obesity. If eating less and moving more didn't permanently cure obesity, perhaps eating *more* and moving *less* wasn't the only cause.

It's something we desperately need to work out. Our species is in the midst of a warbler-like collective weight gain. And just as in warblers, the amount of weight we have gained does not quite tally with changes in 'calories consumed' and 'calories expended'. Even the biggest and most comprehensive of studies show that most of the weight we have gained as a species is not accounted for by the extra food we are eating, nor by our lack of physical activity. Some even indicate that we are eating *less* than we used to, and exercising just as much. The scientific debate about whether gluttony and sloth alone can fully account for the exponential rise in obesity over the past sixty years rumbles quietly on. It is a mere scientific undercurrent lapping at the foundations of research that's seen as more relevant: which diets work best?

At the time of Dhurandhar's frustrations, a mysterious disease was spreading through India's chickens, killing the birds and destroying livelihoods. Dhurandhar's family were friends with a veterinary scientist who was involved in looking for the cause and finding a cure. The culprit was a virus, he told Dhurandhar over dinner, and the birds would die with large livers, shrunken thymus glands and a lot of excess fat. Dhurandhar stopped him. 'The dead chickens are especially fat?' he checked. The vet confirmed it.

Dhurandhar was curious. Animals dying of a viral infection would normally be skinny, not fat. Was it possible that a virus could induce weight gain in chickens? Could this be the explanation behind his patients' difficulties in losing weight? Dhurandhar, excited to know more, set up an experiment. He injected one group of chickens with the virus, and left another group alone. Sure enough, three weeks later, he found that the infected birds were far fatter than the healthy ones. It seemed as if the virus had made them gain weight as they fell ill. Could it be that Dhurandhar's patients, and countless other humans around the world, were also infected with the virus?

What is happening to our species is on such an enormous and unprecedented scale, that in the distant future, when humanity looks back on the twentieth century, they'll remember it not just for two world wars, nor solely for the invention of the internet, but as the age of obesity. Take a human body from 50,000 years ago and one from the 1950s, and they will look more similar to one another than either does to the average human body today. In just sixty years or so, our lean, muscular, hunter-gatherer-like physiques have been encased inside a layer of excess fat. It's something that has never happened to humans on this scale before, and no other animal species – apart from the pets and livestock we care for – has succumbed to this anatomy-changing disease.

One in every three adults on Earth is overweight. One in nine is obese. That's the average across all countries, including those where under-nutrition is more common than being overweight. Looking just at the figures for the fattest of countries is even harder to believe. On the South Pacific island of Nauru, for example, around 70 per cent of adults are obese, and a further 23 per cent are overweight. Just 10,000 people live in this tiny country, and only about 700 of them are a healthy weight. Nauru is officially the fattest nation on Earth, but it is closely followed by most of the other South Pacific islands and several Middle Eastern states.

In the West, we have gone from being skinny enough that no one thought to comment on, worry about, or count the number of overweight people, to being fat enough that it would be quicker to count those that remain skinny. Roughly two in three adults are overweight, and half of those are not just overweight, but obese. The United States, despite its reputation, is seventeenth in the world rankings, with a mere 71 per cent of the population overweight or obese. As for the UK, it ranks thirty-ninth, with 62 per cent of adults overweight (including 25 per cent obese): the highest figures in Western Europe. Even among children in the Western world, being too fat is shockingly common, with up to one-third of under-twenty-year-olds overweight – half of them obese.

Obesity has crept upon us in a way that makes it seem almost normal. Yes, there is a steady stream of articles and news pieces about the obesity 'epidemic' to remind us that it is actually a problem, but we have very quickly adapted to living in a society where most people are overweight. We are quick to assume that fatness is the next step along from greediness and laziness, but if that's the case, it's quite an indictment of human nature. Looking at our other achievements as a species over the past century or so – the inventions of mobile phones, the internet, aeroplanes, life-saving medicines and so on – suggests we are not all just lying around, stuffing ourselves with cake. The fact that lean people are now in the minority in the developed world, and that this change has happened in just fifty or sixty years, after thousands upon thousands of years of human leanness, is shocking – just what are we doing to ourselves?

On average, people in the Western world have gained roughly a fifth of their own body weight in the last fifty years alone. If your allotted time on Earth had fallen so that your 'today' fell in the 1960s, not the 2010s, you would, in all likelihood, be considerably lighter. People who are 11 stone in 2015 might well have been just over 9 stone in 1965, no special efforts required. Today, to regain a pre-1960s weight, tens of millions of people are perpetually on a diet, attempting to deprive themselves of foods for which their brains have a deep-rooted desire. But despite the billions of dollars spent on fad-dieting, gym-going and pill-taking, obesity levels rise inexorably.

This rise has taken place in the face of sixty years of scientific research into effective weight-maintenance and weight-loss strategies. In 1958, back when being overweight was still relatively rare, one of the pioneers of obesity research, Dr Albert Stunkard, said: 'Most obese persons will not stay in treatment for obesity. Of those who stay in treatment, most will not lose weight. And of those who do lose weight, most will regain it.' He was broadly right. Even half a century later, success rates in trials of weight-loss intervention strategies are extremely low. Often, less than half of participants achieve weight loss, and for most it's just a few kilos over a year or more. Why is it so very hard?

Up to now, among those looking for explanations – and perhaps excuses – for their weight, genetics has been the fashionable place to lay the blame. Differences in human DNA, though, have not proved to be particularly illuminating when it comes to weight gain, with only a tiny proportion of our susceptibility to obesity explained by genes. In 2010, a huge study was conducted by a team of hundreds of scientists who hunted through the genes of a quarter of a million people in the hope of finding some that were associated with weight. Astonishingly, they discovered just 32 genes in our 21,000-strong genome that seemed to play a role in weight gain. The average difference in weight between people with the very lowest genetic likelihood of obesity and the very highest was just 8 kg (17 lb). For those who would like to blame their parents, that equates to between 1 per cent and 10 per cent extra risk of becoming overweight, and that's for those people who are in possession of the worst combination of those gene variants.

Regardless of the genes involved, genetics could never be the full explanation for the obesity epidemic, because sixty years ago almost everyone was slim, despite having broadly the same gene variants as the human population today. What probably matters far more is the impact that a changing environment – our diets and lifestyles, for example – has on the workings of our genes.

Our other favourite explanation is that of a 'slow metabolism'. 'I don't have to watch what I eat, I've got a quick metabolism,' must be one of the most irritating comments a lean person can make, but it has no basis in science. A slow metabolism – or more correctly, a low basal metabolic rate – means that a person burns relatively little energy whilst doing absolutely nothing at all – no moving, no watching TV, no doing mental arithmetic. Metabolic rates do vary from person to person, but it is actually overweight people who have the faster metabolisms, not lean people. It simply takes more energy to run a big body than a small one.

So if genetics and low basal metabolic rates aren't behind the obesity epidemic, and the amount we eat and move doesn't fully explain our collective weight gain, what is the explanation? Like many people, Nikhil Dhurandhar wondered if there's more to it than we assume. The possibility that a virus could be causing or exacerbating obesity in some people played on his mind. He tested fifty-two of his human patients in Mumbai for antibodies to the chicken virus – evidence that they had been infected by it at some stage. To his surprise, the ten most obese of these patients had had the virus at some point. Dhurandhar made up his mind; he would stop trying to treat obesity and start researching its causes instead.

We have reached the point in human history where we are considering, in the United Kingdom at least, that redesigning and re-routing the digestive system that evolution has given us is the best way to prevent us from eating ourselves to death. It seems that gastric bands and bypasses, which reduce the size of the stomach and prevent people from consuming everything that their brains and bodies tell them to, are the most effective and the cheapest way to get a grip on the obesity epidemic and its consequences for our collective health.

If diets and exercise are so futile that gastric bypasses are our only hope for significant weight loss, what does that say for the straightforward application of the laws of physics – energy intake minus energy burned equals energy stored – to us as animals governed by the laws of biochemistry?

As we are just beginning to learn, it's not that simple. As the warblers and many hibernating mammals show, there's more to managing weight than counting calories. Subscribing to a simple one-in, one-out system of balancing the body's energy books utterly undermines the great complexities of nutrition, appetite regulation and energy storage. As George Bray, a doctor who has been researching obesity since the start of the epidemic, once said: 'Obesity isn't rocket science. It's much more complicated.'

Two and a half thousand years ago, Hippocrates – the father of modern medicine – believed that all diseases begin in the gut. He knew little of the gut's anatomy, let alone of the 100 trillion microbes that live there, but as we are learning two millennia later, Hippocrates was on to something. Back then, obesity was relatively uncommon, as was another twenty-first-century illness that clearly has its origins in the gut: irritable bowel syndrome. It's with this most unpleasant of maladies that our microbes come into the picture.

In the first week of May 2000, unseasonably heavy rain drenched the rural town of Walkerton, Canada. As the rainstorms passed, Walkerton's residents began to fall ill in their hundreds. With ever more people developing gastroenteritis and bloody diarrhoea, the authorities tested the water supply. They discovered what the water company had been keeping quiet for days: the town's drinking water was contaminated with a deadly strain of *E. coli*.

It transpired that bosses at the water company had known for weeks that the chlorination system on one of the town's wells was broken. During the rains, their negligence had meant that run-off from farmland had carried residues from manure straight into the water supply. A day after the contamination was revealed, three adults and a baby died from their illnesses. Over the next few weeks, three more people succumbed. In total, half of Walkerton's 5,000-strong population were infected in just a couple of weeks

Even though the water supply was quickly cleaned and made safe to drink, the story didn't end there for many of those that had fallen ill. The diarrhoea and cramps just kept coming. A full two years later, one-third of the people affected were still ill. They had developed post-infectious irritable bowel syndrome (IBS), and more than half of them still had it eight years after the outbreak.

As new IBS sufferers, these unfortunate Walkerton residents had joined the growing ranks of people in the West whose bowels rule their lives. For most with the condition, severe abdominal pain and unpredictable bouts of diarrhoea determine the freedom of their days. For others, it's the opposite – constipation, and the pain that goes with it, lasting days and sometimes weeks at a time. 'At least,' says the British gastroenterologist Peter Whorwell of those with constipation-predominant IBS, 'these patients can get out of the house.' For a minority, the double difficulty of both diarrhoea and constipation makes daily life particularly unpredictable.

The trouble is, even though nearly one in five people in the West – mostly women – are stuck with this life-changing illness, we don't actually know what it is. It's not normal, that much is clear. The word 'irritable' belies the impact that IBS has on the lives of its sufferers; the disease is consistently ranked as reducing quality of life even more than for patients on kidney dialysis and diabetics reliant on insulin injections. Perhaps it's the hopelessness that comes with not knowing what's wrong, nor how to fix it.

The spread of IBS is an unremarked global pandemic. One in ten visits to the doctor relate to the condition, and gastroenterologists are kept in a job by the steady flow of sufferers who make up half of their patients. In the United States, IBS leads to 3 million visits to the doctor, 2.2 million prescriptions, and 100,000 hospital visits each year. But we keep it quiet. No one wants to talk about diarrhoea.

The cause, however, remains elusive. Whereas the colon of a person suffering from inflammatory bowel disease would be coated with ulcers, the intestines of people with IBS appear as pink and smooth as those of a healthy person. This lack of physical signs has led to IBS becoming tainted by the historical assumption that it is all in the mind. Though for most sufferers their IBS is at its worst when they are stressed, it's unlikely that stress alone is the sole cause of such a persistent illness. The staggering proportion of people with IBS deserves an explanation – we haven't been through millions of years of evolution just to need to be within thirty seconds of a loo.

A clue is to be found in the Walkerton tragedy. The people stuck with IBS after the water-contamination incident are not the only IBS sufferers to blame their illness on a gastrointestinal infection of some kind. Around a third of patients pinpoint the moment their gut troubles started on an episode of food poisoning or similar, which never seemed to resolve. Traveller's diarrhoea is often the beginning – people who pick up a bug abroad are up to seven times more likely to get IBS. But testing for the original bug yields nothing – they are no longer suffering from the gastroenteritis itself. It's as if the original infection has thrown the gut's normal residents into disarray.

For others, the onset of their IBS coincides not with an infection, but with a course of antibiotics. Diarrhoea is a common side effect of taking certain antibiotics, and for some patients it continues long after all the pills have been popped. There's a paradox though, as antibiotics can also be used to treat IBS, apparently staving off the problem for weeks or months at a time.

So what's going on? These clues – the gastroenteritis and the antibiotics – hint at a common theme: that short-term disruption of the gut's microbes can have long-lasting effects on the microbiota's composition. Imagine a virgin rainforest, verdant and dense with life: insects rule the undergrowth and primates hoot from the canopy. Now see the loggers move in, chainsawing the forest's leafy infrastructure, established over millennia, and bulldozing the rest. Imagine too a weed invading, perhaps having hitchhiked as a seed on the wheels of the diggers, and then crowding out the natives as it takes hold. The forest will regrow, given time, but it will not be the same pristine, complex, unspoilt habitat it was before. Diversity will drop. Sensitive species will die out. Invaders will flourish.

For the complex ecosystem of the gut, on a scale a million times tinier, the principle still stands. Antibiotic chainsaws and invasive pathogens pull apart the web of life that's forged a balance through countless subtle interactions. If the destruction is large enough, the system cannot bounce back. Instead, it collapses. In the rainforest, this is habitat destruction. In the body, it causes *dysbiosis* – an unhealthy balance of microbiota.

Antibiotics and infections are not the only causes of dysbiosis. An unhealthy diet or a nasty medication might have the same effect, throwing off the healthy balance of microbial species and reducing their diversity. It is this dysbiosis, in whatever form it takes, that sits at the heart of twenty-first-century illnesses, both those that start – and end – in the gut, like IBS, and those that affect organs and systems all around the body.

In IBS, the impact of antibiotics and gastroenteritis suggests that the chronic diarrhoea and constipation might be rooted in gut dysbiosis. You can detect what species are living in people's guts and what their abundances are using DNA sequencing. Doing that with people with IBS and healthy people shows that most people with IBS have distinctly different microbiotas than people without it. Some IBS patients, though, have microbiotas that are no different from those of healthy people. These patients tend to report being depressed, suggesting that for a small subset of IBS sufferers, psychological illness drives the IBS, whereas for others, dysbiosis is the primary cause, and stress simply worsens it.

Amongst IBS sufferers with dysbiosis, some research has found differences in the composition of their microbiotas according to the type of IBS they have. Patients who complained of being bloated and feeling full quickly when eating had higher levels of *Cyanobacteria*, whereas those who suffered a lot of pain had a greater quantity of *Proteobacteria*. For constipated patients, a whole community of seventeen bacterial groups were present in the gut in increased numbers. Other studies have found that not only is an altered microbiota present, but it is highly unstable compared with that of healthy people, with different groups of bacteria waxing and waning over time.

In retrospect, it might seem predictable that irritable bowel syndrome is likely to be a consequence of bowels 'irritated' by the 'wrong' microbes. As a logical extension it is highly plausible: from a quick bout of diarrhoea brought on by bacteria in dirty water or undercooked chicken, to chronic bowel dysfunction, all because the gut's bacteria have got out of balance. But whereas a diarrhoeal illness can often be blamed on a particular pathogenic bacterium – for example, *Campylobacter jejuni* in the case of food poisoning from raw chicken – IBS can't be pinned on one nasty bug. Instead, it seems to be something about the relative numbers of what are normally seen as 'friendly bacteria'. Perhaps not enough of one variety, or too many of another. Or even a species that behaves itself under normal circumstances but turns nasty given a chance to take over.

If the gut community found in IBS patients has no overtly infectious player, how exactly does dysbiosis wreak such havoc with the functioning of the gut? The groups of bacteria present in the gut of a person with IBS also seem to be present in the gut of a healthy person, so how can changes in their numbers alone be responsible? At the moment, this is proving a difficult question for medical scientists to answer, but studies have revealed some interesting clues. Although IBS sufferers do not have ulcers on the surface of their intestines as in inflammatory bowel disease, their guts are more inflamed than they should be. It's likely the body is attempting to flush the microbes out of the gut, by opening tiny gaps between the cells lining the gut wall and allowing water to rush in.

It's easy to imagine how having the wrong balance of microbes in the gut could cause IBS. But what about gut trouble of a different kind – the expansion of the human waistline? Could the microbiota be the missing link between calories-in and calories-out?

Sweden is a country that takes obesity very seriously. Although ranked as only the ninetieth-fattest country on the planet, and one of the slimmest in Europe, Sweden has the highest rate of gastric bypass surgeries in the world. The Swedish have considered implementing a 'fat tax' on high-calorie foods, and doctors are able to prescribe exercise to overweight patients. Sweden is also home to a man who has made one of the biggest contributions to forwarding obesity science since the epidemic began.

Fredrik Bäckhed is a professor of microbiology at Gothenburg University, although it's not Petri dishes and microscopes you'll find in his lab, but dozens of mice. Like humans, mice play host to an impressive collection of microbes, mainly living in their guts. But Bäckhed's mice are different. Born by Caesarean section and then housed in sterile chambers, they do not have any microbes in them. Each one is a blank canvas – 'germ-free', which means that Bäckhed's team can colonise them with whichever microbes they wish.

Back in 2004, Bäckhed took a job with the world's leading expert on the microbiota, Jeffrey Gordon, a professor at Washington University in St Louis, Missouri. Gordon had noticed that his germ-free mice were particularly skinny, and he and Bäckhed wondered if this was because they lacked gut microbes. Together, they realised that even the most basic studies on what microbes did to an animal's metabolism had not yet been done. So Bäckhed's first question was simple: Do gut microbes make mice gain weight?

To answer that question, Bäckhed reared some germ-free mice to adulthood, and then dotted their fur with the contents of the caecum – the chamber-like first part of the large intestine – of mice who had been born normally. Once the germ-free mice had licked the caecal material off their fur, their guts took on a set of microbes like any other mouse. Then something extraordinary happened: they gained weight. Not just a little bit, but a 60 per cent increase in body weight in fourteen days. And they were eating less.

It seemed it wasn't only the microbes that were benefiting from being given a home inside the mice's guts, but the mice as well. Everybody knew that microbes living in the gut were eating the indigestible parts of the diet, but no one had ever looked into how much this second round of digestion contributed to energy intake. With microbes helping them to access more of the calories in their diets, the mice could get by on less food. In terms of our understanding of nutrition, it really rocked the boat. If the microbiota determined how many calories mice could extract from their food, did that mean they might be involved in obesity?

Microbiologist Ruth Ley – another member of Jeffrey Gordon's lab group – wondered if the microbes in obese animals might be different to those in lean animals. To find out, she used a genetically obese breed of mice known as *ob/ob*. At three times the weight of a normal mouse, these obese mice look nearly spherical, and they just will not stop eating. Although they appear to be a completely different species of mouse, they actually have just a single mutation in their DNA that makes them eat non-stop and become profoundly fat. The mutation is in the gene that makes leptin, a hormone which dampens the appetite of both mice and men if they have a decent supply of stored fat. Without leptin informing their brains that they are well-fed, the *ob/ob* mice are literally insatiable.

By decoding the DNA sequences of the barcode-like 16S rRNA gene of the bacteria living in the guts of the *ob/ob* mice, and working out which species were present, Ley was able to compare the microbiotas of obese and lean mice. In both types of mice, two groups of bacteria were dominant: the Bacteroidetes and the Firmicutes. But in the obese mice, there was half the abundance of Bacteroidetes than in the lean mice, and the Firmicutes were making up the numbers.

Ley, excited by the possibility that this difference in the ratio of Firmicutes to Bacteroidetes might prove to be fundamental to obesity, then checked the microbiotas of lean and obese humans. She found the same ratio – the obese people had far more Firmicutes and the lean people had a greater proportion of Bacteroidetes. It seemed almost too simple – could obesity and the composition of the gut microbiota be connected in such a straightforward way? Most importantly, were the microbes in obese mice and humans *causing* the obesity, or were they just a consequence of it?

It fell to a third member of Gordon's lab group to find out – PhD student Peter Turnbaugh. Turnbaugh used the same kind of genetically obese mice as Ley had, but he transferred their microbes into germ-free mice. At the same time, he transferred the microbes of normal, lean mice into a second set of germ-free mice. Both sets of mice were given exactly the same amount of food, but fourteen days later, the mice colonised with the 'obese' microbiota had got fat, and those with the 'lean' microbiota had not.

Turnbaugh's experiment showed not only that gut microbes could make mice fat, but also that they could be passed between individuals. The implications go far beyond moving bacteria from obese mouse to lean mouse. We could be doing it the other way round – taking microbes from lean people and putting them into obese people – weight loss with no dieting required. The therapeutic – and money-making – potential was not lost on Turnbaugh or his collaborators, who have patented the concept of altering the microbiota as a treatment for obesity.

But before we get too excited about the potential for a cure for obesity, we need to know how it all works. What are these microbes doing that makes us fat? Just as before, the microbiotas in Turnbaugh's obese mice contained more Firmicutes and fewer Bacteroidetes, and they somehow seemed to enable the mice to extract more energy from their food. This detail undermines one of the core tenets of the obesity equation. Counting 'calories-in' is not as simple as keeping track of what a person eats. More accurately, it is the energy content of what a person *absorbs*. Turnbaugh calculated that the mice with the obese microbiota were collecting 2 per cent more calories from their food. For every 100 calories the lean mice extracted, the obese mice squeezed out 102.

Not much, perhaps, but over the course of a year or more, it adds up. Let's take a woman of average height, 5 foot 4 inches, who weighs 62 kg (9 st 11 lb) and has a healthy Body Mass Index (BMI: weight (kg) /(height (m)^2) of 23.5. She consumes 2,000 calories per day, but with an 'obese' microbiota, her extra 2 per cent calorie extraction adds 40 more calories each day. Without expending extra energy, those further 40 calories per day *should* translate, in theory at least, to a 1.9 kg weight gain over a year. In ten years, that's 19 kg, taking her weight to 81 kg (12 st 11 lb) and her BMI to an obese 30.7. All because of just 2 per cent extra calories extracted from her food by her gut bacteria.

Turnbaugh's experiment has set in motion a revolution in our understanding of human nutrition. The calorie contents of foods are normally calculated using standard conversion tables, so every gram of carbohydrate is deemed to contribute 4 calories, every gram of fat, 9 calories, and so on. These labels present the calories of a food item as a fixed value. They are saying: 'This yogurt contains 137 calories', and 'A slice of this bread contains 69 calories.' However, Peter Turnbaugh's work suggests it is not that straightforward. That yogurt may well contain 137 calories for a person of normal weight, but it could also contain 140 calories for someone who is overweight, and who has a different set of gut microbes. Again, it's a small difference, but it adds up.

If your microbes are working on your behalf to extract energy from your food, it is your particular community of microbes that determines how many calories you get from what you eat, not a standard conversion table. For those people who have dieted without success, this may be part of the explanation. A carefully calculated calorie-controlled diet, resulting in an overall loss of calories every day for a sustained period, should lead to weight loss. But if the 'calories-in' are underestimated, that could mean no change in weight, or even weight gain. This idea is backed up by another experiment carried out by Reiner Jumpertz in 2011 at the National Institute of Health in Phoenix, Arizona. Jumpertz gave human volunteers a fixed-calorie diet, and simply measured the calories that remained in their stool after digestion. Lean volunteers put on a high-calorie diet had a boost in the abundance of Firmicutes relative to Bacteroidetes. This change in gut microbes went along with a drop in the number of calories that were coming out in their stool. With the balance of bacteria shifted, they were extracting an extra 150 calories each day from the same diet.

The particular set of microbes we harbour determines our ability to extract energy from our food. After the small intestine has digested and absorbed as much as it can from what we've eaten, the leftovers move into the large intestine, where most of our microbes live. Here, they function like factory workers, each breaking down its own preferred molecules and absorbing what it can. The rest is left in a simple enough form for us to absorb through the lining of the large intestine. One strain of bacteria might have the genes needed to break down the amino-acid molecules that come from meat. Another strain might be best suited to breaking down the long-chain carbohydrate molecules that come from green vegetables. And a third could be most efficient at collecting up the sugar molecules that were not absorbed in the small intestine. The diet each of us eats affects which strains we harbour. So, for example, a vegetarian might not have many individuals of the amino-acid strain, as they can't proliferate without a steady supply of meat.

Bäckhed suggests that what we can extract from our food depends on what our microbial factory has been set up to expect. If our vegetarian were to abandon her stance and indulge in a hog roast, she would probably not have enough amino-acid-loving microbes to make the most of it. But a regular meat-eater would have a sizeable collection of suitable microbes, and would extract more calories from the hog roast than the vegetarian. And so it follows for other nutrients. A person who eats very little fat would have very few microbes that are specialised for fat, and the odd doughnut or chocolate bar might make it through the large intestine without being efficiently stripped of its remaining calorific content. Someone who eats a daily tea-time treat, however, would have a large population of fat-munching bacteria, just waiting to strip their next doughnut to its bare essentials, providing our snacker with the full dose of calories.

Although the number of calories we absorb from our food is undoubtedly important, it's not just how much energy our microbes extract for us that matters, but what they make the body do with that energy. Do we use it immediately to power our muscles and our organs? Or do we store it for later, in case there's nothing to eat? Which of these things happens depends on our genes. But it's not which gene variants you got from your parents that matters, it's which genes are switched on and which are switched off, which are dialled up and which are dialled down.

Our own bodies do the turning on and off of genes, and the dialling up and down, using all sorts of chemical messengers. This control means that cells in our eyes can do different jobs than cells in our livers, for example. Or that cells in the brain can function differently when we're working during the day than when we're sound asleep in the middle of the night. But our bodies are not the only masters of our genetic output. Our microbes also get a say, controlling some of our genes to suit their needs.

Members of the microbiota are able to turn up production from genes which encourage energy to be packed away in our fat cells. And why not? The microbiota benefits from living in a human who can make it through the winter just as much as the human does. An 'obese microbiota' turns up these genes even more, forcing the storage of extra energy from our food as fat. Annoying as this may be for those of us who struggle to maintain the weight they'd like, this gene-control trick should be beneficial, as it helps us to make the most of our food and store that energy away for leaner times. In our past, in periods of feast and famine, having help to get through the famines would have been a life-saver.

Calories-in, then, goes deeper than what you put in your mouth. It's what your gut absorbs, including what your microbes provide for you. Calories-out, too, is more complicated than how much energy you use being active. It's also about what your body chooses to do with that energy: whether it stores it away for a rainy day, or burns it off immediately. Although both of these mechanisms show how one person might absorb and store more than another, depending on the microbes they host, it raises another question: why don't people who absorb more energy and store more fat simply feel satiated sooner? Why, if they have absorbed plenty of calories, and stored plenty of fat, are some people driven to keep on eating?

Your appetite is governed by many things, from the immediate, physical sensation of a full stomach to hormones that tell the brain how much energy is stored as fat. The chemical I mentioned earlier that was missing in the genetically obese mice – leptin – is one such hormone. It is produced directly by fat tissue, so the more fat cells we have, the more leptin gets released into the blood. It's a great system – it tells the brain we're satiated once we have accumulated a healthy amount of stored fat, and our appetites are suppressed.

So why don't people lose interest in food once they start to put on weight? When leptin was discovered in the 1990s, courtesy of the *ob/ob* mice who were genetically unable to make leptin of their own, there was a flurry of excitement about using the hormone to treat patients with obesity. Injecting *ob/ob* mice with it led to very rapid loss – they ate less, they moved around more, and they dropped to nearly half of their body weight in a month. Even giving leptin to normal, lean mice made them lose weight. If mice could be treated this way, could leptin be the cure for human obesity?

The answer, as is obvious from the continuing obesity epidemic, was no. Giving obese people leptin injections had hardly any effect on their weight or their appetites. Though disappointing, this failure shed light on the true nature of obesity. Unlike in the *ob/ob* mice, it is not too little leptin that allows people to become fat. In fact, overweight people have particularly high levels of leptin, because they have extra fat tissue that produces it. The trouble is, their brains have become resistant to its effects. In a lean person, gaining a bit of weight leads to extra leptin production, and a decrease in appetite. But in an obese person, though plenty of leptin is being produced, the brain can't detect it and so they never feel full.

This leptin resistance hints at something important. In obesity, normal mechanisms of appetite regulation and energy storage have fundamentally changed. Excess fat is not just a place to pack away unburnt calories, it's an energy-usage control centre, a bit like a thermostat. When the body's fat cells are comfortably full, the thermostat clicks off, reducing appetite and preventing further food intake from being stored. Then as fat stores fall low, the thermostat clicks back on again, increasing appetite and storing more food as fat. As in the garden warblers, weight gain is not just about eating more, it's about biochemical shifts in how the body manages energy. This 'warbler effect' undermines the basic assumption that balancing what we eat with how much we move is all that's necessary to maintain a stable weight. If this belief is wrong, perhaps obesity is not a straightforward 'lifestyle disease', born of gluttony and sloth, but an illness with an organic origin beyond our control.

If this seems like too radical a suggestion, consider this: just a few decades ago, stomach ulcers were 'known' to be caused by stress and caffeine. Like obesity, they were thought of as a lifestyle disease

– change your habits and the problem will go away. The solution was simple: keep calm and drink water. But this treatment didn't work – patients returned again and again, with acid burning holes in their stomachs. The cause of their failure to recover was deemed obvious; these patients must not be sticking to their treatment plans, allowing stress to prevent them from getting better.

But then, in 1982, two Australian scientists, Robin Warren and Barry Marshall, discovered the truth. A bacterium called *Helicobacter pylori* that sometimes colonised the stomach was causing ulcers and the related condition gastritis. Stress and caffeine simply made them more painful. Such was the resistance of the scientific community to Warren and Marshall's idea, that Marshall drank a solution of *H. pylori*, giving himself gastritis in the process, to prove the connection. It took fifteen years for the medical community to fully embrace this new cause. Now, antibiotics are a cheap and effective way to cure ulcers for good. In 2005, Warren and Marshall won a Nobel Prize in Physiology or Medicine for their discovery that stomach ulcers were not a lifestyle disease, as dogma dictated, but the result of an infection.

Likewise, with his virus, Nikhil Dhurandhar was challenging the dogma that obesity was a lifestyle disease – one of excess. To investigate the possibility that a viral infection could be capable of causing weight gain in humans, he needed to switch from practising medicine to researching the science behind it. Dhurandhar decided to uproot his family and move to the United States, in the hope of getting the research funding he would need to find answers to his questions. It was a leap of faith, in the face of stiff opposition from the scientific establishment. But it would eventually pay off.

Two years after he had moved to America, Dhurandhar had still not managed to convince anyone to support his research into the chicken virus. He was on the brink of giving up and returning to India when the nutritional scientist Professor Richard Atkinson, at the University of Wisconsin, agreed to give him a job. At last, Dhurandhar was ready to begin his experiments. But there was a major obstacle:

the American authorities refused them permission to import the chicken virus into the United States – it might cause obesity, after all.

Together, Atkinson and Dhurandhar devised a new plan. They would study another virus – this time one that was common in Americans – in the hope that it too might be responsible for weight gain. Based on a hunch that it was similar to the chicken virus, they selected a different virus that was known to cause respiratory infections from a laboratory catalogue, and ordered it through the post. Its name was Adenovirus 36, or Ad-36.

Once again, Dhurandhar began his experiments with a group of chickens. He infected half the birds with Ad-36 and the other half with a different adenovirus, more normally found in birds. And then he and Atkinson waited. Would Ad-36 make the birds fat, just as the Indian virus had done?

If it did, Dhurandhar would be making a big claim. He would be suggesting that overeating and being under-active were not the sole cause of human obesity; that the obesity epidemic might have another origin; that obesity could be an infectious disease – not just a lack of will-power. And most controversially, Dhurandhar would be implying that obesity was contagious.

Looking at maps charting the spread of the obesity epidemic in the United States over the past thirty-five years certainly gives the impression that it's an infectious disease sweeping through the population. The epidemic begins in the south-eastern states and quickly begins to spread. As more and more people become overweight at the epicentre of the disease, it pushes outward to the north and west, affecting ever-greater swathes of the country. Hot spots crop up in major cities, setting off new bubbles of obesity that expand over time. Though a handful of scientific studies have commented on this infectious-disease-like pattern, it's usually put down to the spread of an 'obesogenic environment' – more fast-food restaurants, super-markets with calorie-dense foods, and lifestyles with ever less physical activity.

One study in people has found that obesity spreads in a manner that mimics a contagious disease even on an individual level. By analysing the weights and social connections of over 12,000 people over thirty-two years, researchers found that a person's chances of becoming obese are closely tied to weight gain in their nearest and dearest. For example, if a person's spouse became obese, that person's risk of becoming obese themselves went up by 37 per cent. Fair enough, you might think – they probably have the same diet. But the same is true for adult siblings, most of whom don't live together. More strikingly, if a person's close friend became obese, then that person's risk of following suit shot up by 171 per cent. It didn't seem to be down to choosing friends of a similar weight – these people were close before the weight gain. Neighbours who were not counted as friends were exempt from the increased risk of obesity, which makes it seem less likely to be the opening of a fast-food restaurant or closure of a gym nearby that pushes the tied weight gain among social groups.

Of course, there are many possible social reasons for this phenomenon – a shared shift in attitude towards obesity, or joint consumption of unhealthy foods, for example – but a thought-provoking addition to the list of explanations is that of microbial cross-over, viral or otherwise. Even if Dhurandhar's virus is not the main culprit, there are many other microbes to consider. Perhaps the sharing of 'obesogenic' members of the microbiota within social networks contributes to the obesogenic environment that other researchers perceive, and facilitates the spread of obesity. Closer friends are more likely to spend time in one another's homes, sharing surfaces, food, bathrooms – and microbes. Sharing such microbes may allow obesity to spread that little bit more easily.

The final day of Dhurandhar's experiment with the chickens arrived. For him, the results justified the sacrifices he and his family had made in leaving their lives and loved ones behind in India. Just as with the other virus, Ad-36 had made the infected chickens grow fat, whilst those given the alternative virus had remained lean. Dhurandhar was finally able to publish his discovery in the scientific literature, but many more questions lay ahead. Most importantly, did Ad-36 work the same way in humans? Could a virus be making people fat?

Dhurandhar and Atkinson knew they could not intentionally infect humans with the virus – if it *did* make people gain weight, they would have no way of curing them again. Instead, they had to do the next best thing and test the virus in another primate species – a small monkey known as a marmoset. As with the chickens, the infected marmosets gained weight. Dhurandhar thought he was on to something big. To see whether the virus was at least associated with obesity in humans, he decided to screen the blood of hundreds of volunteers for antibodies to Ad-36. Sure enough, 30 per cent of the obese volunteers had had the virus, compared to just 11 per cent of the lean volunteers.

Ad-36 is a case-in-point when it comes to the warbler effect. Having the virus doesn't make chickens eat more or move less, it makes their bodies store more of the energy in their food as fat. Like the bacteria living in an obese person, Ad-36 is meddling with the normal energy-storage system. Exactly how big a contribution this virus has made to the obesity epidemic remains unknown, but, as with the story of the warblers, it tells us something important: obesity is not always a lifestyle disease caused by overeating and being under-active. Rather, it is a dysfunction of the body's energy-storage system.

In theory, it's possible to calculate *exactly* how much weight a person should gain from a particular number of excess calories in their diet, just as I did a few pages back. For every 3,500 calories we consume beyond our energy needs, we should gain 1 lb of fat. We could eat that excess in a single day, or over an entire year, but the result should be the same – we become 1 lb heavier in that day, or over that year.

But in practice, it doesn't work like this. Even in some of the earliest studies of weight gain, the numbers didn't add up. In one experiment, researchers fed twelve pairs of identical twin men an excess of 1,000 calories per day, six days a week for 100 days. In total, each man ate 84,000 calories more than their bodies needed. According to theory, this should have led to a weight gain of exactly 24 lb in every man. In reality, it was not that straightforward. For starters, even the *average* amount the men gained was far less than maths dictates that it should have been, at 18 lb. But the individual gains betray the real failings of applying a mathematical rule to weight gain or loss. The man who gained the least managed only 9 lb – just over a third of the predicted amount. And the twin who gained the most put on 29 lb – even more than expected. These values aren't '24 lb, more or less', they are so far wide of the mark that using it even as a guide is purposeless.

The mere fact that weight gain can deviate so dramatically from that predicted by weighing up calories-in against calories-out shows the warbler effect is not limited to migratory birds and hibernating mammals. Ultimately, the laws of thermodynamics can't be overruled – energy in *must* equal energy out for weight to remain stable. But the crucial point is this: mechanisms within the body that go beyond simply how much we eat and how much we move are responsible for regulating both the calories we absorb and, even more so, the calories we either expend or store.

Ad-36 provides a good example of one way this can work. Fat tissue sitting under the skin and around the organs is made of cells that sit empty, waiting to be filled with fat when it's time to store energy. In chickens infected with Ad-36, the virus forces these cells to fill up, even when there's not a large excess of energy for storage. The chickens didn't need to eat more to get fatter – their bodies were just pushing energy into storage rather than using it elsewhere.

So is something like this going on in human obesity? Are obese people storing fat in a different way from lean people? Patrice Cani, Professor of Nutrition and Metabolism at the Université catholique de Louvain in Belgium, knew that not only were obese people resistant to the effects of the satiety hormone leptin, they also showed signs of illness within their fat tissue. Unlike in lean people, their fat cells were flooded with immune cells, almost as if they were fighting an infection.

Cani also knew that when lean people stored energy, they made *more* fat cells, filling each with just a small amount of fat. But in obese people, this healthy process of energy storage wasn't happening. Instead of making more fat cells, they were making *bigger* fat cells, stuffing them with more and more fat. To Cani, the inflammation and the lack of new fat cells were a sign that overweight people had gone beyond a healthy process of energy storage and into an unhealthy state. This wasn't the kind of weight gain that would help people survive a hard winter any more. It was, in itself, a kind of illness, he thought.

Cani suspected that it was the 'obese' microbiota that were causing the inflammation and the shift in fat storage. He knew that some bacteria living in the gut were coated in a molecule called lipopolysaccharide, or LPS, which acted like a toxin if it got into the blood. Sure enough, Cani learnt that obese people had high levels of LPS in the blood. It was the LPS that was responsible for triggering the inflammation in their fat cells. Even more tellingly, Cani discovered that the LPS was preventing new fat cells from forming, so existing fat cells were being overfilled.

It was a great leap forward. The fat of obese people was not just layer upon layer of stored energy, it was fat tissue that had biochemically malfunctioned, and LPS seemed to be causing that malfunction to occur. But how was the LPS getting from the gut into the blood?

Among the microbes that are present in different amounts in the guts of lean and obese people is a species called *Akkermansia muciniphila*. This bacterium correlates neatly with weight – the less *Akkermansia* a person has, the higher their BMI. About 4 per cent of the microbial community of lean people belong to this species, but obese people have barely any. It lives, as its name suggests, on the surface of the thick layer of mucus that coats the gut lining (*muciniphila* means mucus-loving). This mucus forms a barrier that prevents the microbiota from crossing into the blood, where they could turn nasty. The amount of *Akkermansia* people have is not only linked to their BMI. The lower the abundance of this bacterium, the thinner a person's mucus layer, and the more LPS they have in their blood.

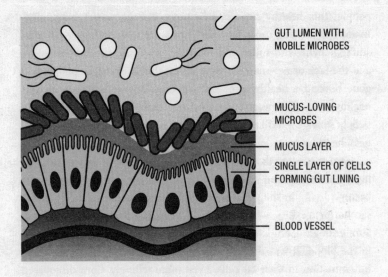

GUT LUMEN WITH
MOBILE MICROBES

MUCUS-LOVING
MICROBES

MUCUS LAYER

SINGLE LAYER OF CELLS
FORMING GUT LINING

BLOOD VESSEL

The gut lining.

It might seem like *Akkermansia* is common in lean people's guts because it is reaping the benefits of their thick mucus layers, but it is actually responsible for persuading the cells of the gut lining to produce more mucus. *Akkermansia* sends chemical requests which dial up the human genes that make mucus, thereby providing itself with a home, and preventing LPS from crossing into the blood.

If this bacterium could boost the thickness of the mucus layer, Cani thought, perhaps it could reduce LPS levels and prevent weight gain too. He tried supplementing the diets of a group of mice with *Akkermansia*, and sure enough, their LPS levels dropped, their fat tissue began creating healthy new cells again, and, most importantly, they lost weight. The mice given *Akkermansia* had also become more sensitive to leptin, meaning their appetites decreased. The weight the mice had gained was not because they ate too much, it was because the LPS was forcing their bodies to store energy rather than spend it. As Dhurandhar suspects, this warbler-like shift in energy storage suggests that people may not always be obese because they eat too much.

Sometimes – perhaps more often than not – they may eat too much because they're ill.

Cani's discovery that *Akkermansia* protects mice against obesity could prove to be revolutionary. He is planning to test its effects in overweight humans in the hope that it could be provided as a supplement to fight weight gain. Ultimately, we still need to know what is causing the drop in numbers of *Akkermansia* in overweight and obese people. There are a few clues: making mice obese by putting them on a high-fat diet lowers *Akkermansia* levels, but supplementing that diet with fibre brings numbers of the bacterium back up to healthy levels again.

By 2030, 86 per cent of the US population are predicted to be either overweight or obese. By 2048, everyone will be. We have spent fifty years attempting to tackle obesity through encouraging people to eat less and move more. It hasn't worked. Millions more adults and children become overweight or obese every year despite the vast sums of money we spend trying to stay slim or lose weight. Yet we continue to approach the treatment of obesity just as we did half a century ago, with precious little progress.

Now, our only consistently effective treatment for obesity is gastric-bypass surgery. Patients who have tried and failed to lose weight by dieting are deemed to need their stomachs reduced to the size of an egg to stop them from overeating. The assumption is that they couldn't stick to a diet, so they must take drastic action to keep their caloric intake down. Within weeks of the surgery, patients will lose several stone.

It is an intervention that was assumed to work by removing the need for will-power, physically preventing people from eating more than a child's portion at each meal. But it seems there might be more to it than portion control. Within a week of bypass surgery, the gut microbiota stop looking like that of an obese person, and start looking like that of a lean person. The Firmicutes-to-Bacteroidetes ratio reverses, and good old *Akkermansia* becomes around 10,000 times as common. Performing tiny gastric-bypass surgeries in mice shows the same changes in the composition of the mouse-gut microbiota. But sham surgeries, in which incisions are made but the stomach is re-sewn in its original place, don't have the same effect. Even transferring the microbiota from a mouse that's had bypass surgery into a germ-free mouse causes sudden weight loss. It seems to be the re-routing of nutrients, enzymes and hormones that causes the shift in species which then leads to weight loss. It seems it is not the tiny portions that patients are limited to that helps them lose weight, but the change in energy regulation courtesy of their new 'lean' microbiota.

Twenty-five years after the outbreak of the chicken virus in Mumbai, Nikhil Dhurandhar was made President of the Obesity Society in the United States. His research into viral causes of obesity has steadily gained scientific acceptance, and he continues to investigate the underlying causes of obesity, beyond the calories-in and calories-out that we see on the surface of the epidemic.

Microbes, both viral and bacterial, are showing us that there's more to obesity than eating too much and moving too little. The energy each of us extracts from our food, and the way in which that energy is used and stored, is intricately linked with the particular community of microbes we host. If we really want to get to the heart of the obesity epidemic, we need to look inward to the microbiota and ask what we are doing to alter the dynamic that they established with the human body in its leanest, healthiest form.

THREE

Mind Control

Every now and then, in the pesticide-laced wetlands of the western states of North America, frogs and toads appear with grotesque malformations. Many have up to eight hind limbs, splaying out from the hips, and some are missing legs altogether. They struggle to swim and hop, and are often picked off by birds before they are fully grown. This developmental abnormality is not a genetic mutation, but the work of a microbe – a parasitic trematode worm. The larvae of this worm, having been expelled from their previous host, the ram's horn snail, seek out the frogs when they are tadpoles. They burrow into the limb buds and form cysts, which disrupt the normal development of the legs, sometimes causing them to duplicate, then duplicate again.

For the frogs, their deformation is usually fatal, as they struggle to escape from hungry herons looking for an easy meal. For the trematode, the extra limbs help their life cycle to continue. A heron can easily catch the frogs and digest them along with their resident trematodes, unwittingly becoming their next host. Soon enough, they are returned to the water in the heron's droppings, where they take up again in the ram's horn snail. A clever strategy, it seems – not that intelligence plays any part in the trematode's journey through its hosts. It is natural selection that brings the frogs their misfortune, as the trematodes that create frogs vulnerable to predation are those that survive, and can pass on their limb-deforming, life-cycle-perpetuating genes.

Altering your host's body is one way to improve your evolutionary fitness – your chance of reproducing – but there is another: altering your host's behaviour.

Turning over leaves at about head height in the rainforests of Papua New Guinea occasionally reveals the husks of dead ants, their jaws clamped to the central veins, holding their lifeless bodies in place. Out of each ant sprouts a long stalk, bent over with the weight of its spore-filled sac. These stalks are *Cordyceps* fungi, which have killed the ants and are feeding off their bodies, releasing spores that fall to the forest floor. Growing within ants is a clever means of gaining the energy needed to reproduce, but for the fungi the ants serve a further purpose.

Once infected by *Cordyceps*, an ant becomes a zombie. It ignores its usual duties with its colony on the forest floor, and succumbs to the urge to climb a tree. Here it finds a leaf vein on the northern side of the trunk, about 150 cm above the forest floor, and bites down hard, anchoring itself in place. This is known as the 'death grip', as shortly afterwards, the fungus takes the ant's life. Days later, the fungus germinates, sprouts a stalk and releases its spores. They drift down, blanketing the leaf litter below and infecting a new troop of ants. The *Cordyceps* has changed the ant's behaviour, controlling it in an incredibly specific way, to help it give rise to the next generation of fungi.

It's not the only behaviour-changing microbe. Dogs infected by rabies, far from curling up to die, turn extremely aggressive. Frothing at the mouth with virus-filled saliva, rabid dogs go looking for a fight, desperate to bite another dog. Rats infected by the *Toxoplasma* parasite lose their fears of open spaces and bright light. They become drawn to the scent of bobcat urine, effectively seeking out their main predator. Insects infected by a parasitic hairworm appear to commit suicide by jumping into water, where the hairworm is able to emerge from the dead insect.

Each of these microbe-controlled behaviours has evolved because it helps the microbe to spread to new hosts. Rabies viruses that make dogs bite are passed on to another dog or two, where they continue to reproduce. *Toxoplasma* parasites that are able to steer rats towards cats can continue their life cycle when the rat is killed and eaten. Hairworms must reach a water source to find mates and reproduce. Being able to control the behaviour of the host makes the microbes that can do it more capable of surviving and reproducing, and so evolution favours them. What's astounding is just how precise this control can be.

The effects of microbes on behaviour are not limited to wildlife. Humans, too, can be exposed to microbial whimsy. Take the case of a Belgian girl, Miss A, who had, up to the age of eighteen, been happy and healthy, and working towards her college exams. Over the course of a few days, she became aggressive, refused to communicate and lost her sexual inhibitions. She was sent to a psychiatric hospital, where she was prescribed anti-psychotic medication and discharged. Three months later, Miss A returned to the hospital with worsening behaviour and uncontrollable vomiting and diarrhoea. Her doctors decided to take a biopsy of her brain, which revealed the source of her psychiatric condition – a microbe. She had Whipple's disease – a rare infection caused by a bacterium which occasionally announces its presence through the behaviour of its host.

It's interesting to note that Miss A had gastrointestinal symptoms – the vomiting and diarrhoea – alongside the behavioural symptoms that landed her in a psychiatric hospital. People suffering from Whipple's disease normally present to their doctors with rapid weight loss, abdominal pain and diarrhoea – all signs of a gastrointestinal infection. In Miss A, that infection had influenced not only her gut, but also her brain, turning her doctors' attention away from its true source. In fact, gastrointestinal symptoms are surprisingly common in people with mental health and neurological conditions, though they are usually seen as unimportant compared with the altered behaviour.

For one extraordinary woman, though, her autistic son's diarrhoea proved to be a clue worth paying attention to.

Ellen Bolte was already the mother of three children when her fourth, Andrew, was born in February 1992 in Bridgeport, Connecticut. He was, as her daughter Erin and her other two sons had been, a happy and healthy baby, meeting all the expected developmental milestones. At the time of his fifteen-month check-up with the paediatrician, Andrew seemed well – his usual self. But to Ellen's surprise, the doctor was horrified by the state of Andrew's ears. They were full of fluid, he said – Andrew had a bad ear infection and needed antibiotics. 'I was surprised, because he hadn't had a fever and he was eating, drinking and playing as normal,' Ellen said. But after a ten-day course, when Ellen returned for a follow-up, the fluid had not shifted. A second ten-day course was prescribed, this time using a different antibiotic. At the end of the course, Andrew's ears had cleared.

But this proved temporary and the saga continued. Andrew was prescribed a third, and then a fourth course of antibiotics to try to clear his ears for good – all different medications targeting different groups of bacteria. At this stage, Ellen began to question the need for more drugs, as her son didn't seem to have any discomfort or hearing trouble. But the doctor was insistent – 'If you value your son's hearing, you will give him these antibiotics,' he told her. Ellen relented and did as she was told. That's when her son's diarrhoea began. But diarrhoea is often a side effect of antibiotics, so rather than stopping them, the doctor prescribed yet another thirty-day course to keep the infection at bay.

During this final course, Andrew's behaviour changed. At first he seemed as if he were a little drunk, grinning and staggering. 'Like a happy drunk,' said Ellen, 'I joked with my husband that the next time we had a party we could spike the punch with his antibiotics, to liven things up. We thought maybe Andy's ears had been causing him such pain that he was now delirious with happiness to be pain-free.' But that was short-lived, and after a week Andrew began to pull away. He became withdrawn and sullen, and then intensely irritable, screaming all day long. 'Before the antibiotics, I didn't have a sick child. *Now* I had a very sick child.' More gastrointestinal (GI) symptoms cropped

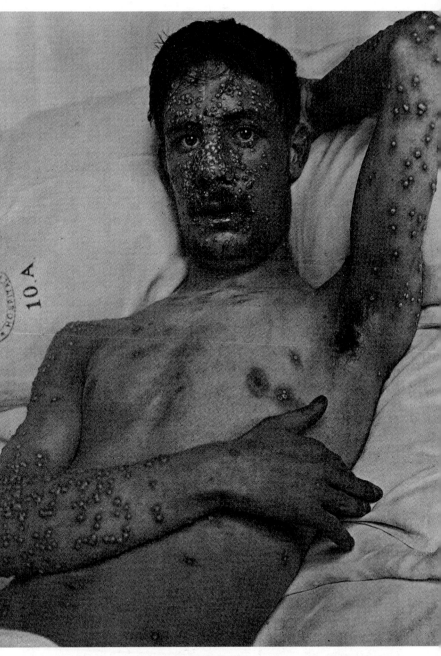

Smallpox was common even in developing countries until the 1900s.
This man was photographed around the turn of the century.

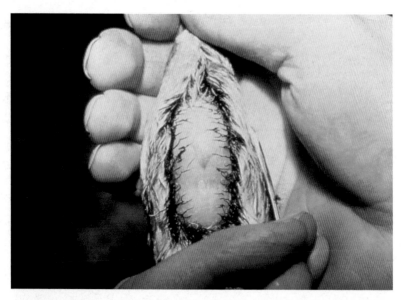

Parting the sparse feathers on the underside of a garden warbler (above) reveals the large energy reserves stored as fat following binge eating and a metabolic shift in preparation for migration, in comparison to a lean garden warbler (below) without these fat reserves.

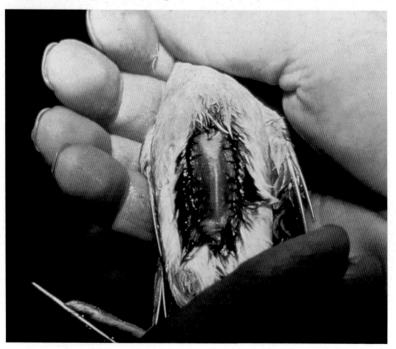

A mutation in the gene for the hormone leptin leads to genetically obese (*ob/ob*) mice, which are around three times the weight of their lean littermates.

Adult Obesity Trends in the United States (BMI ≥30)

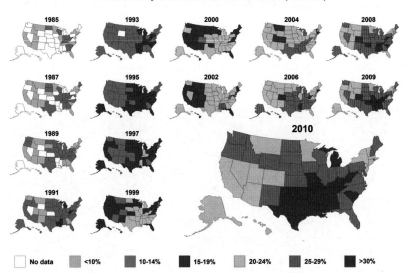

The prevalence of obesity (defined as a BMI of 30kg/m² or more) has increased significantly over time, as shown by these maps of data collected by America's Centers for Disease Control's Behavioral Risk Factor Surveillance System. In 2012, 35 per cent of American adults and 17 per cent of children were obese. A further 34 per cent of adults and 15 per cent of children were overweight, but not obese.

Ellen Bolte with Andrew and his older sister Erin on Christmas Day 1993, shortly after Andrew developed autism.

Despite lacking a background in biology, Ellen Bolte was compelled to investigate the possible microbiological causes of autism in her son Andrew as a toddler. The Bolte family in 2011 – (from left to right) Andrew, Ellen, Erin and Ron.

An ant in Papua New Guinea that has succumbed to the behaviour-controlling fungus, *Cordyceps*, which compels it to climb a tree and bite into the central vein of a leaf before it dies, scattering fungal spores over the forest floor.

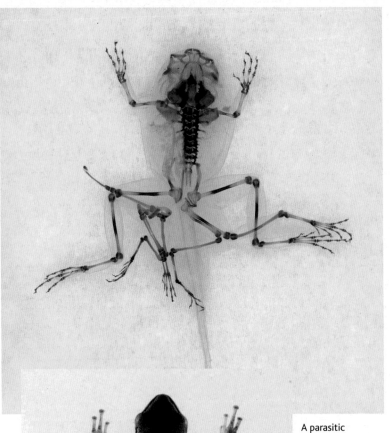

A parasitic trematode worm has evolved to cause limb abnormalities in American frogs, easing its transmission into its next host, the heron, where it continues its life cycle. Its efforts are supported by the contamination of wetlands with pesticides.

Male greater sac-winged bats entice females by wafting a microbially-refined perfume over them as they roost.

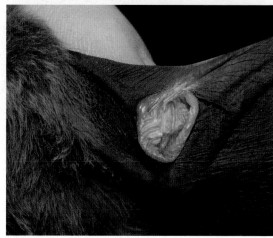

The perfume is carefully blended by allowing a precise community of microbes to ferment a mixture of urine, saliva and semen within specialised pouches on the wings.

Using Robogut at the University of Guelph, Canada, graduate student Erin Bolte is testing her mother Ellen's hypothesis that the microbes of the gut can be responsible for autism.

up – Andrew's diarrhoea worsened and was full of mucus and undigested food.

Andrew's behaviour deteriorated. 'He began to do all these really
bizarre behaviours – walking on the tips of his toes and avoiding looking at me. The few words that he had, he lost,' Ellen said. 'He wouldn't
even respond when I'd call his name. It was as if he was gone.' Ellen
and her husband took Andrew to an ear specialist, who inserted tiny
tubes to help his ears drain. The specialist said there was no ear infection, and recommended they took Andrew off cow's milk. At this
point, his ears cleared. Ellen was hopeful: 'I thought, OK, now that
we've got his ears cleared up, his behaviour will return to normal. But
very quickly, it became clear that that wasn't happening.'

By now, Andrew's GI symptoms were dreadful, and despite having
been a healthy weight, he became skinny with a very bloated belly. His
behaviour also became more bizarre. He walked up on his toes without
bending his knees. He stood in the doorway, flipping the light switch
on and off for half an hour at a time. He was preoccupied by objects,
such as pots with lids, but not interested in other children. Most of
all, he screamed. At this point, Andrew's parents, desperate for help,
took him from doctor to doctor looking for answers. At twenty-five
months old, he was diagnosed with autism.

For many people, including Ellen Bolte at the time of Andrew's
diagnosis, the 1988 movie *Rain Man*, in which Dustin Hoffman plays
an autistic character, is their only impression of autism. In the film,
despite severe difficulties with social interaction and insistence on
his daily routine, Hoffman's character has an extraordinary memory and is able to recall years of data about the American baseball
league. He is an autistic savant – both challenged and gifted. Despite
the media interest in it, the remarkable musical, mathematical and
artistic abilities of savantism are rare. In reality, autism forms a
spectrum of symptoms, from those with average or above-average
intelligence – known as Asperger syndrome – to those with severe
autism and significant learning disabilities like Andrew Bolte.

In common to all with autistic spectrum disorder (ASD) are

difficulties with social behaviour. It was this feature that prompted the American psychiatrist Leo Kanner to identify autism as a distinct syndrome in 1943. In his seminal paper on the topic, he described the case histories of eleven children who shared an '*inability to relate themselves* in the ordinary way to people and situations from the beginning of life'. Kanner borrowed the word *autism*, meaning 'self-ism', from the constellation of symptoms associated with schizophrenia. 'There is from the start,' he wrote, 'an *extreme autistic aloneness* that, whenever possible, disregards, ignores, shuts out anything that comes to the child from the outside.' Sufferers can struggle to understand tone and intent, perhaps missing the joke or taking sarcasm and metaphors literally. They may have difficulty empathising with other people, and understanding the unwritten social rules the rest of us pick up throughout childhood. And autistic people might prefer a fixed routine, or become obsessively focused on a single idea or object.

In the 1990s, when Andrew Bolte was diagnosed, it was thought that all children with autism were born with it, as Leo Kanner had noted. For Ellen, this meant that diagnosis *had* to be wrong. 'I knew, with every ounce of my being, that Andrew hadn't been born with it. I'd had four kids – he was just fine, absolutely fine.' But despite her protestations, the doctors would only maintain that she must have missed the signs – that Andrew had been autistic for as long as he had been alive. Her conviction that he had not been, and therefore that he was not autistic at all, meant the correct diagnosis was still out there, waiting to be uncovered. It was this conviction that pushed Ellen into the research that would ultimately lead her to a game-changing hypothesis of autism's cause.

Autism was once exceedingly rare, probably affecting around 1 in 10,000 people. By the time the first real surveys were done in the late 1960s, around 1 in 2,500 children were affected. In the year 2000, America's Centers for Disease Control and Prevention began keeping records, at first recording that autistic spectrum disorder affected 1 in 150 eight-year-old children. This figure rose rapidly over the following decade, hitting 1 in 125 in 2004, 1 in 110 by 2006 and 1 in 88 by 2008. At the time of the last count, in 2010, the number stood at 1 in every 68 children, having more than doubled in the past ten years.

Plotted on a graph, these figures look extremely worrying, as the rise shows few signs of letting up. Following the trend into the future leads to a picture of a very different society. Even conservative estimates suggest that 1 in every 30 children could be autistic by 2020, and some have suggested that every family in America will have a child on the autistic spectrum by 2050. ASD tends to affect boys more often than girls, such that more than 2 per cent of boys are now considered to be on the autistic spectrum. Although some say that better diagnosis has led to a false rise, and though increased awareness has undoubtedly contributed to today's figures, the experts agree that the rise in autism cases is genuine. Until recently though, few could agree on the cause.

At the time Ellen Bolte began her research, the prevailing theory of the cause of autism was a genetic one. Only a decade previously, a great many psychiatrists believed in the 'refrigerator mother' hypothesis, which had been inadvertently put forward by Leo Kanner in 1949. He wrote that autistic children were exposed from the beginning of their lives to 'parental coldness, obsessiveness, and a mechanical type of attention to material needs only . . . They were left neatly in refrigerators which did not defrost. Their withdrawal seems to be an act of turning away from such a situation to seek comfort in solitude.' But Kanner also wrote that autism was an innate disorder, beginning before birth, and has since said that he therefore never believed that parents were responsible for making their children unwell. By the 1990s, though the refrigerator mother idea persisted in some parts of the world, it had largely been refuted and attention had turned, somewhat fashionably perhaps, to the role of genetics in causing autism.

Of course, Ellen was not meaning to solve the riddle of the autism's cause. Instead, she was looking into the possibility that a sudden event – an illness or exposure of some kind – might have made her son ill. Ellen began, with the perfect balance of open-mindedness and scepticism belonging to the best of scientists, to search for answers. Her career as a computer programmer stood her in good stead, as she worked through a series of logical steps to piece together a hypothesis. She began, despite having had no training in medicine or science, at the beginning: with observation. 'I watched him, and I thought about what was making him behave the way he did. He would eat ash from the fireplace and tissue paper, despite refusing the food I gave him. What was making him do that? He reacted as if he was in pain when he was touched, or when there were loud noises. Again, why?'

Starting with the public library, Ellen read everything she could that might provide a clue. She continued to see doctors, looking for an alternative diagnosis, or just someone interested enough not to turn her and Andrew away. One doctor took an interest in Andrew's case, and told her that if she really wanted to do research, she needed to start reading the medical literature. Intimidating though this was, Ellen picked up speed and became familiar with medical jargon. After several false starts, she began to focus on whether the antibiotics Andrew had received for his ears might have caused the damage. She stumbled upon the emerging research on *Clostridium difficile* infections, which cause severe intractable diarrhoea in some people after they've been given antibiotic treatment. The link with Andrew's GI symptoms jumped out at her, and Ellen wondered whether a similar bacterium might not only cause diarrhoea, but also release a 'toxin' that could have affected Andrew's developing brain.

At this point, Ellen hit upon her hypothesis: she thought Andrew must be infected with a bacterium related to *Clostridium difficile*, called *Clostridium tetani*. But instead of getting into the blood and causing a tetanus infection in the muscles, as it normally does, Ellen suspected *C. tetani* had got into his gut. She had a hunch that the antibiotics Andrew had received for his ear infection had killed the protective bacteria living in his gut, allowing *C. tetani* to take hold. From there, she thought, the neurotoxin these bacteria produce had somehow travelled to Andrew's brain. Ellen was excited, and took the idea to her doctor.

'He was very open-minded. He said anything that we could reasonably test for, we would do the test.' They screened Andrew's blood for evidence that his immune system had dealt with a *C. tetani* infection. Like most small children in America, Andrew had been immunised against tetanus, so evidence in his blood of some immune protection to it was inevitable. But the blood test result shocked even the laboratory staff – Andrew's level of immune protection was off the chart; nothing like that seen in children after immunisation. After months of blood tests coming back negative, Ellen Bolte finally had some evidence that she was on the right track.

Ellen began writing to doctors, asking them to consider her theory and requesting that they treat Andrew with a further antibiotic, vancomycin, to rid him of the *C. tetani* in his gut. Doctor by doctor, Ellen's ideas were rebutted. Why didn't Andrew have extreme muscle contraction, as is typical of tetanus? How had the neurotoxin got through the blood–brain barrier and into the brain? How could Andrew be infected with a bacterium he had been immunised against? But Ellen was convinced, from months of research, that she was right.

With each doctor's dismissal, Ellen delved further into the scientific literature, looking for answers to their questions. She found out that muscle contractions occurred after infection through a wound in the skin, which allowed the neurotoxin to seek out nerves to the muscles, rather than an infection in the gut where it could affect nerves to the brain. She learnt about experiments in which the path of the tetanus neurotoxin had been traced from gut to brain via the vagus nerve – a major connection between the two organs – providing a route around the blood–brain barrier. She dug up case histories of patients who had contracted classic tetanus infections despite having been adequately immunised. Over time, Ellen had come to accept Andrew's diagnosis of autism. Her search had turned from a very personal quest into a fresh perspective on a disease without any apparent cause.

By the time Ellen approached the thirty-seventh doctor on her list, she was thoroughly briefed in every aspect of her hypothesis. Dr

Richard Sandler, a paediatric gastroenterologist at the Rush Children's Hospital in Chicago, spent two hours listening to Ellen tell Andrew's tale and put forward her ideas. At the end, he asked for two weeks to think over her proposal to treat Andrew with yet more antibiotics – this time to kill the *C. tetani*. 'As absurd as it sounded,' he said, 'it was scientifically *plausible*. I couldn't dismiss it.'

Dr Sandler agreed to go ahead with an eight-week trial of antibiotics for Andrew, who was by now four and a half years old. Prior to the treatment, he ran a battery of blood, urine and stool tests and also brought in a clinical psychologist who did a series of observations of Andrew's behaviour, so that they could measure any changes that might occur during the treatment period. A few days after Andrew began taking the antibiotics, he became even more hyperactive than usual. What came next would astonish Dr Sandler, justify Ellen's two-year battle with the medical establishment, and, eventually, change the face of autism research.

Ever the observer, Charles Darwin noted in his 1872 book *The Expression of the Emotions in Man and Animals* that: 'The manner in which the secretions of the alimentary canal . . . are affected by strong emotions is another excellent instance of the direct action of the sensorium on this organ, independently of the will.' He refers, of course, to the bowel-loosening sensation that accompanies the delivery of bad news, or the stomach-turning feeling attached to the realisation you've slept through your alarm and are late for an exam, and even the butterflies-in-the-tummy giddiness of falling in love. The brain and the gut, despite their physical distance and thoroughly different functions, share an intimate connection. It goes both ways: not only do emotions affect the workings of the gut, but the activity of the gut can affect mood and behaviour. Think of the last time you suffered an upset tummy – no doubt it was not only your digestive system feeling grumpy, but you as well.

For sufferers of long-term conditions such as irritable bowel syndrome, emotions play a fearsome role in their symptoms. When stress levels are high, IBS can play up, making a stressful situation even more difficult. The emotion of a first date or important presentation at work is only heightened by the discomfort and worry of IBS, bringing on a vicious cycle of worsening symptoms and stress. Knowing that IBS is linked to changes in the gut microbiota, is it possible that the gut–brain connection features a third player? Should we think of it as a gut–*microbiota*–brain connection instead?

The Japanese medical scientists Nobuyuki Sudo and Yoichi Chida were the first to ask this question, back in 2004. They designed a simple experiment, using mice, that looked at whether the gut microbiota affected the brain's response to stress. They used two groups of mice: in one group, the mice were germ-free – no microbes were living in their guts – but the other group of mice had a normal set of gut microbes. When they stressed the mice by placing them in a tube, both groups produced stress hormones, but in the germ-free mice the hormone concentrations were twice as high. Without their microbiota, the mice had found the situation much more stressful.

Sudo and Chida wanted to know whether they could reverse the overreaction to stress in the germ-free mice by colonising them with a normal microbiota as adults. As it turned out, it was too late – their stress response was already fixed. It emerged that the earlier the mice were colonised, the less stressed they became. Surprisingly, colonising the germ-free mice as youngsters with even just one single bacterial species, *Bifidobacterium infantis*, was enough to prevent them becoming any more stressed than the mice with a normal gut microbiota.

These experiments opened the door to a new way of thinking. Gut microbes did not just alter physical health, but mental health as well. What's more, it seemed that the effects could begin in childhood if the gut microbiota were disturbed early on. Our brains go through an intensely concentrated period of development when we are babies and toddlers. At birth, we each have nearly our full allotment of around 100 billion nerve cells – neurons– in our brains. But these are just the raw materials, like a pile of wooden planks. Building anything meaningful requires some careful joinery work using connections, called synapses, that bind the neurons together. A young child's experiences dictate the synapses that are formed, as well as which are important enough to be reinforced, and which are insignificant enough to discard. A toddler, whose daily life is packed full of novel stimulation, will form around 2 million synapses a second, each offering new potential for learning and development. Healthy brains need a careful balance between remembering and forgetting, so most of these new synapses will be discarded during childhood. As they say, it's 'use it or lose it' with synapses at that age – any that are not regularly reinforced will be pruned to leave a nice tidy brain.

If gut microbes can influence this crucial period of brain development in early life, could this support Ellen Bolte's idea that her son Andrew's autism was due to a gut infection? Regressive-onset autism affects children before the age of three, during the same period as the bulk of brain development. This time-frame also coincides with the establishment of a stable, adult-like gut microbiota. Andrew's early antibiotic treatment for what appeared to be an ear infection would have disrupted that process, potentially allowing a takeover by the neurotoxin-producer *Clostridium tetani*. Ellen hoped that giving Andrew further antibiotics might destroy the *C. tetani* she believed had infected him, halting the damage being done to his young brain.

Two days of profound calm followed Andrew's hyperactivity after beginning his treatment. 'It was a miracle,' Ellen said. 'We got a few weeks into the treatment and the light bulbs just started to come on. I started him on potty training – he was four years old! – and was able to complete that in a few weeks. He was understanding what I was saying to him for the first time in three years.' Andrew became affectionate, responsive and calm, and even learnt to speak more than the few words he had picked up before he became ill. He would allow himself to be dressed, and what's more, he would not have eaten his way down the front of his T-shirt by the end of the day. The child psychologist prepared a report on Andrew's behaviour during the antibiotic treatment, but Dr Sandler hardly needed one, so profound were the changes.

Spectacular though this improvement was, a sample size of one – Andrew – could never prove beyond doubt that the source of autism lay in the gut. Fortunately, a microbiologist of formidable reputation had been brought in on Ellen's ideas after the success of Andrew's antibiotic trial. Dr Sydney Finegold had devoted his career to the huge subset of 'anaerobic' bacteria that exist without oxygen. He was referred to, in a paper written in celebration of his ninetieth birthday in 2012, as 'arguably the most influential investigator in anaerobic microbiology in the twentieth century or perhaps for all time'. It was to this group of bacteria that the members of the genus *Clostridium*, including *C. tetani*, belonged. With Finegold's scientific calibre and knowledge, Ellen's autism hypothesis was in good hands.

Dr Sandler, together with Finegold and Ellen, extended the antibiotic trial to eleven other children with delayed-onset autism and diarrhoea. The aim was not to see if antibiotics might make a suitable treatment for autism, but rather to use it as a proof-of-concept. If these drugs could make these children even partially or temporarily better, then the microbes living in the gut, *C. tetani* or otherwise, might be responsible. As with Andrew, the result of the trial in the other children was dramatic. They began making eye contact, playing normally and using language to express themselves. They became less obsessed with a single object or activity and they were more agreeable. Sadly, for neither Andrew nor the other children did the improvements in health and behaviour during the trial last. Within a week or so of stopping the antibiotics, most of the children had regressed to their previous state. But for the first time since the illness was described, the mystery of autism had a new, and promising, lead: the microbes of the gut.

In 2001, six years after Ellen Bolte had first developed her hypothesis that *C. tetani* in the gut was the cause of autism, she would at last find out if she had been right. Sydney Finegold organised a study of the microbes living in the large intestines of thirteen children with autism and eight healthy 'control' children for comparison. DNA-sequencing techniques were still prohibitively expensive for a full survey of the children's microbiota, but Finegold's skill at culturing bacteria in oxygen-free conditions meant that the species belonging to the genus *Clostridium* could be counted. Though *C. tetani* itself was not found, something was not right. Compared with the healthy children, the autistic children had, on average, ten times the number of clostridia in their guts. Perhaps, like *C. tetani*, these related species were also producing a neurotoxin that could damage the brains of young children. Ellen Bolte's hypothesis had not quite hit the nail on the head, but it seemed at this stage that she had only missed by a species' breadth.

Is it really possible that simply having a different collection of bacteria in the gut could make children flap their hands, rock back and forth, and scream for hours, as many autistic children do? Quite possibly. It turns out that the *Toxoplasma* parasite – the one that makes rats lose their fear of open spaces and become attracted to the scent of cat urine – also changes the behaviour of humans. We're prone to becoming infected because of our love of cats – even domestic cats carry the parasite, and it's easy to catch it from a scratch or from the litter tray. So easy, in fact, that a massive 84 per cent of Parisian women tested for it during pregnancy were found to be infected. Elsewhere the figure tends to be a bit lower: around 32 per cent of pregnant women in New York City and 22 per cent in London for example. For a developing baby, a fresh *Toxoplasma* infection can be very dangerous, hence the testing of pregnant women, but in the adult population it rarely causes ill-health. The parasite does leave its mark though, in the form of an altered personality.

Strangely enough, *Toxoplasma* infection has almost the opposite effect in men and women. Infected men tend to become less pleasant, disregarding societal rules and losing their sense of morality. They are, relatively speaking, more suspicious, jealous and insecure. For women though, the effects of infection seem almost desirable, as they become more easygoing, warm-hearted and trusting. They also tend to be more self-assured and decisive than uninfected women. It's intriguing to think of the potential for promiscuity here, with females having dropped their defences and males acting with less regard for others and newly loosened morals. Fundamentally, like the rats, humans of both sexes appear to be more open to risk: women through an increase in trust, and men through social thoughtlessness.

Personality change is not the only effect of *Toxoplasma* infection in people. Both men and women show slower responses and are more likely to lose concentration once infected. The effects of this, though mild in laboratory tests, can have serious consequences. By comparing the frequency of *Toxoplasma* infection found in 150 people hospitalised in Prague after they caused road traffic accidents, with that of citizens who had not caused an accident, a team of researchers at the Charles University calculated that being infected makes you nearly three times as likely to cause an accident. A similar study in Turkey found that drivers involved in accidents were four times as likely to be infected.

Unlike rats, humans are dead-end hosts for the *Toxoplasma* parasite, as our chances of being eaten by a bobcat tend to be pretty slim. But through the lingering effect of our evolutionary history, where death-by-feline was perhaps as likely as death-by-car is now, this little parasite can shift our personality and change our behaviour. An alternative explanation is that the parasite was never meant for us, and instead the mechanism that *Toxoplasma* evolved to get itself from rat to cat simply works just as well on the human brain as on the rodent's. Either way, it's enough to make you wonder whether your own body plays host to this meddling little creature, and which personal calamities you might be able to attribute to it.

Aside from the somewhat amusing shifts in personality that *Toxoplasma* can cause, infection with this parasite also has a darker side. Way back in 1896, *Scientific American* magazine published an article entitled 'Is Insanity Due to a Microbe?' At the time, the idea that microbes could cause diseases at all was a new one, so it was natural that the concept might be extended to psychiatric conditions. A couple of doctors from a hospital in New York State had tried injecting the spinal fluid of patients suffering from schizophrenia into rabbits, which subsequently fell ill, making them wonder what microbes might be lurking in mentally ill patients.

Though lacking in scientific rigour, this mini-experiment spurred a flurry of interest in microbes as the cause of mental health problems. Despite the great promise of the idea, the work of Sigmund Freud led to it being unceremoniously dropped a few decades later in favour of his emerging psychoanalytic theory. Instead of a physiological cause for neurological conditions, Freud proposed an emotional one, rooted in the experiences of childhood. His theory held fast until lithium turned out to be a better cure for manic depression than talking.

As disease after disease was found to be caused by microbes in the first half of the twentieth century, the peculiar maladies of just one organ were deemed to be exempt from microbial influence: the brain. Given the futility of talking the kidneys out of failing, or the heart out of stopping, it's amazing to think of the effort that has gone into curing the brain of its ailments primarily through discussion. When any other organ breaks down, we look for external causes, but when the brain – the mind! – misbehaves, we assume it's the fault of the individual, their parents or their lifestyle.

Perhaps because of its special place in our sense of self and free will, the brain did not receive the scrutiny of microbiologists again until the final years of the twentieth century. At this point, many microbes were soon linked to mental illness, but it is the *Toxoplasma* parasite that has proved to be the most compelling suspect for many conditions. Occasionally, when people are first infected with the parasite they develop psychiatric symptoms, such as hallucinations and delusions, that lead to an initial misdiagnosis of schizophrenia. In fact, amongst those with schizophrenia, the presence of *Toxoplasma* is three times more common than in the general population – a far more telling association than any genetic connections so far revealed.

Intriguingly, schizophrenics are not the only mental health patients in whom *Toxoplasma* infection is rife. It has also been found to be involved in obsessive–compulsive disorder (OCD), attention deficit hyperactivity disorder (ADHD) and Tourette's syndrome, all of which have become increasingly common over the past several decades. The old idea that psychiatric conditions can be caused by microbes is back, but this time it has a new, and more subtle, twist. Instead of known enemies like *Toxoplasma* getting all the blame, what if our resident microbes are also stirring up trouble?

If members of the microbiota are really capable of influencing behaviour, could transplanting the gut microbes cause a personality shift, much as transplanting the gut microbes of an obese mouse can make a lean mouse gain weight? Of course, mice can't fill in personality questionnaires, but different breeds can have characteristic behaviours, just as cat and dog breeds do. One laboratory mouse breed, known as BALB, stands out as being particularly timid and hesitant – completely the opposite of a confident and gregarious breed known as the Swiss mouse. As different as fat and thin, these mice made the perfect partners for a personality swap experiment in 2011.

A team of scientists at McMaster University in Ontario, Canada, discovered that altering the gut microbiota of mice by giving them antibiotics made them less anxious about exploring a new environment. This got them thinking – could they transfer anxiety from the anxious BALB mice into the relaxed Swiss mice by transplanting BALB gut microbes? They inoculated the two breeds with their opposite's bugs, and gave all the mice a simple test. They were placed on a platform inside a box and timed until they worked up the courage to step down and explore. The normally brave Swiss mice took three times longer to step down after receiving the microbes from the anxious mice than they did with their own microbes. Likewise, the nervous BALB mice grew braver and stepped down faster after receiving the Swiss mouse microbiota.

If the nature-over-nurture idea that your personality is not your own hard-earned creation but a product of your genes makes you feel uneasy, how about the concept of a personality composed by the bacteria living in your gut? Mice without gut microbes are antisocial, preferring to spend time alone rather than with other mice. Whereas a mouse with a normal microbiota will choose to meet and greet any new mice added to its cage, germ-free mice stick with mice they already know. Simply having gut microbes seems to make them more friendly. Beyond friendship, it's possible that your microbiota may even affect who you are attracted to.

There is a group of bats in Central America that appear to have a gash on the top of each wing, near the shoulder. These are not wounds, but little pouches, earning this group of species the name 'sac-winged bats'. Male sac-winged bats make good use of their pouches, filling them with bodily secretions: urine, saliva, and even semen. They tend to this potion very carefully, cleaning it out each afternoon and re-filling it to ensure it smells exactly as they want it to. Then, when the time is right, they hover in front of a cluster of females as they hang in the roost, and gently waft the fragrant contents towards them. The effect is, as you might expect, alluring.

The perfect perfume is, it seems, cultivated by getting the right bacterial mix. Each male keeps one or two strains of bacteria in his wing sacs, apparently having 'chosen' them from a selection of about twenty-five species available to male bats. These bacteria feed off the urine, saliva and semen in the wing pouches, and release as waste products a heady mix of sex pheromones, which persuades female bats to join the male's mating harem.

The particular kind of pheromones an animal produces seems to matter even in animals that don't have a special pouch for mixing their own blend of love potion. Fruit flies, whose bodies are little bigger than a pinhead, are nonetheless choosy when it comes to mating. Twenty-five years ago, an evolutionary biologist named Diane Dodd was investigating whether keeping two populations of a species apart would turn them into two different species by changing their ability to breed with one another. She split a group of fruit flies into two, and raised the groups on two different food sources – maltose and starch – for twenty-five generations. Then she reintroduced them, but the flies from the different groups refused to mate. 'Starch' flies would mate with other 'starch' flies, and 'maltose' flies with other 'maltose' flies, but they wouldn't mix their preferences.

It wasn't clear why at the time, but in 2010, Gil Sharon at Tel Aviv University had an idea about what might have caused this response. He repeated Dodd's experiment, and got the same result – the flies were repulsed by one another after just two generations of eating differently. What was causing this change of preference? Sharon suspected that the different foods were altering the gut microbiota of the fruit flies, and changing the smell of their sex pheromones. He then gave them antibiotics to kill off their microbiota, and, sure enough, they no longer cared who they mated with. Without a microbiota, the flies could not produce a characteristic smell. Inoculating them once more with the microbiota from one of the two dietary groups even restored them to their earlier choosy behaviour.

Before you accuse me of over-extrapolating from flies to humans, I'll put this into perspective. The flies' microbiota (actually a single species – *Lactobacillus plantarum*) had apparently altered chemicals that cover the surface of their bodies – essentially sex pheromones. Humans, too, are influenced by sex pheromones. In a now legendary experiment, female students at the University of Bern were given T-shirts worn by male students in bed, and asked to rate them in order of attractiveness. The women preferred the T-shirts of men who had the most different type of immune system from their own. The theory goes that by choosing their genetic opposite, the girls would be providing their offspring with an immune system that can handle twice as many challenges. Through their sense of smell, the girls were scanning the boys' genomes, looking for the best match to be a father to their children.

The odours left by the male students on their T-shirts were produced by none other than the skin microbiota. Living under the armpits, these microbes convert sweat into smells that float off, for better or for worse. Indeed, the sweatiness of the armpits and the groin, and probably the hair that sprouts there, are unlikely to be a cooling mechanism. Rather, they are the human equivalent of the sac-winged bats' perfume pouches, trapping and concocting the perfect scent. The particular community of skin microbiota that each male student harboured is probably at least partly the result of his genes, if mice are anything to go by, including the genes that determine the kind of immune system he has. The girls, albeit unconsciously, were using the microbiota as a messaging system, alerting them to the most beneficial genetic union.

The mind boggles to think of the sheer numbers of ruinous relationships that might have been brought on by deodorants and antibiotics, not to mention contraceptive hormones. In the T-shirt experiment, female students who were taking the pill apparently reversed their subliminal superpower, as they preferred the T-shirts of men whose immune systems were most *similar* to their own.

If microbially-influenced sex pheromones provide a first step in the mate-screening process, consider kissing the next chemical assessment. It might seem like a uniquely human idea – perhaps a cultural phenomenon used to display possession to onlookers without getting too animalistic about it – but in fact we are not the only species to lock lips with one another. Chimpanzees, other primates and many other animals do so too, which gives the purpose of kissing more of a biological slant.

It seems like a pretty risky business, exchanging saliva and germs for the sake of making a connection, especially given that mouth-to-mouth kissing and the touching of tongues is almost exclusively carried out between non-relatives who could be suffering from who-knows-what. But perhaps that's exactly the point. It's a good idea to find out what bugs the potential father of your offspring might be carrying before you make yourself, and your children-to-be, even more vulnerable to those bugs. Not only that, but kissing gives you another, more in-depth, sample of one another's microbiotas. And with that comes a taste of each other's underlying genes and immune capabilities. In kissing each other, we are deciding in whom to place our trust, both emotionally and biologically.

As odd as the idea of microbially-influenced behaviour seems, it raises the possibility of self-improvement by biological means. There's no need for expensive psychiatrists who expect you to delve into the murky depths of your childhood disappointments, if microbes can save you all that money and misery. In a French clinical trial, fifty-five normal, healthy volunteers – humans this time – were given either a fruity-tasting bar containing two strains of live bacteria, or a matching bar but without the bacteria (a placebo). After a month of eating one bar daily, the volunteers given the live bacteria scored happier, less anxious and less angry than they had been before the trial – and the changes went beyond the placebo effect. As trials go, it's short and small, but it offers a glimpse of research avenues worth exploring.

How can eating live bacteria make you feel happier? Pleasingly, one potential mechanism seems to have to do with a chemical that's well known to be involved in mood regulation: serotonin. This neurotransmitter is actually mainly found in the gut, where it keeps everything moving along nicely. But around 10 per cent of serotonin belongs in the brain, regulating mood and even memory. How straightforward it would be if the ingested bacteria simply set up shop in the gut and started pumping out serotonin! But of course, it is not that simple. Instead, introducing live bacteria increases levels of another compound, tryptophan, in the blood. This little molecule is of paramount importance for happiness, as it is converted directly into serotonin. Indeed, depressed patients tend to have lower tryptophan levels in their blood, and countries where the population as a whole eats a diet lower in tryptophan (which is found in protein) have higher rates of suicide. It is even possible to make people profoundly, though temporarily, depressed by depleting their bodies' tryptophan supplies. Less tryptophan means less serotonin, and less serotonin means less happiness.

What's fascinating, though, is that the increase in tryptophan that the added bacteria bring about is not because they make it, but rather because they prevent the immune system from destroying the body's supplies of it. This points to an extraordinary idea – one that has gained ground not only among microbiologists, but in other fields as well. It is becoming increasingly clear that just as with allergies and obesity, depression can be brought on by a malfunctioning immune system. But I'll get back to that.

First I want to tell you about another mechanism by which bacteria can make you happy. It involves the vagus nerve – a major nerve that originates in the brain and extends all the way down to the gut, branching off at various organs along the way. Nerves are like electrical wires – they carry tiny electrical impulses along their lengths that give instructions or sense changes. In the case of the vagus nerve, the impulses carry information about what the gut is doing – what's being digested, how active it is, and so on. The special thing about the vagus nerve, though, is that it informs the brain about what we call 'gut feelings'. Butterflies in the stomach, knowing 'in your gut' that something is wrong, and bowel-loosening nervousness really do begin in the gut – the brain simply gets informed of them by electrical impulses shooting up the vagus nerve.

So perhaps it shouldn't come as a surprise that electrical impulses shooting up the vagus nerve to the brain can also make you happy. Doctors can even treat patients with severe depression – incurable by chemical or behavioural means – by hijacking this system. In this treatment, called vagal nerve stimulation, a tiny device is implanted into the neck of the patient. Extending from this device are wires that the surgeon carefully wraps around the vagus nerve. A battery-powered generator inserted in the chest provides an electrical impulse that stimulates the nerve. Over weeks, months and years, patients steadily become more cheerful, buoyed by their happiness pacemakers.

Adding this electrical pacemaker to the vagus nerve can provide the necessary boost to nerve activity and mood. But under normal circumstances, those electrical impulses have a chemical origin – much like a household battery. These chemicals which kick-start nerve impulses are called neurotransmitters, and you'll have heard of more of them than you might have guessed. Substances like serotonin, adrenalin, dopamine, epinephrine and oxytocin are mostly synthesised by our own bodies, and they are able to initiate a tiny electrical spark at the end of a nerve. But neurotransmitters are not only produced by human cells. The microbiota play their part, producing chemicals that act in the same way, stimulating the vagus nerve and communicating with the brain. The microbes which produce these substances act as a natural vagal nerve stimulator, sending electrical impulses up the vagus nerve and boosting mood. Quite why they have this effect on mood is not clear, but that they do is not in doubt.

One idea is that by influencing our mood, our microbiota can control our behaviour such that it benefits them. Imagine, for example, one strain of bacterium that feeds on a particular compound found in our food. If we eat that food, thus feeding these bacteria, and they are able to 'reward' us with a dose of happiness through the chemicals they produce, so much the better for them. The chemicals they produce in us could cause us to crave the food they feed on, and even to remember where we found it. This would make us return to that location – perhaps a fruit tree in our evolutionary past, or a particular bakery today – making us eat more, and consequently boosting that strain of bacterium, producing further chemicals and extra cravings.

Back to the emerging impact of the immune system on the brain. When the body's armed forces are put on high alert in anticipation of an attack, rogue bullets in the form of chemical messengers called cytokines whizz around, sometimes causing unnecessary damage. These cytokines get the soldiers of the immune system all worked up and primed to fight, but if there's no enemy, friendly fire is all that's left. Depression doesn't seem to be the only neurological outcome of this immune bellicosity. People suffering with many of the other mental health disorders I've mentioned already also show signs of immune overactivity, known as inflammation. ADHD, OCD, bipolar disorder, schizophrenia and even Parkinson's disease and dementia appear to involve an immune overreaction. Adding beneficial bacteria to the gut, as with the people in the French clinical trial, has a calming effect on the immune system. Not only can it prevent the destruction of tryptophan and increase happiness levels, but it can also reduce inflammation.

In autistic patients too, the immune system is busy working away, sending out cytokines to ramp up the aggression levels. Some perceived threat from the altered microbiota in the gut appears to be the catalyst for this, but the question to be answered now is *How?*

Sydney Finegold's studies of the differences between the microbiotas of autistic and healthy children were followed by a number of other attempts to lay blame on particular species. He has even named one bacterial suspect that's more common in autistic children after Ellen: *Clostridium bolteae*. Certainly, the balance of bacteria is different, and often clostridia crop up as being among the culprits. But what are these bacteria doing that so profoundly alters the brains of severely autistic children?

At the University of Western Ontario in London, Canada, is a man whose education and experience has set him up perfectly to play his role on this new scientific stage where the brain and the gut meet. Dr Derrick MacFabe began his medical training in neuroscience and psychiatry. During high school, he had worked with children with special needs, many of whom were autistic and had gastrointestinal (GI) problems. Later, whilst working as a qualified doctor at a hospital, MacFabe encountered patients with GI troubles who were deemed crazy or neurotic, and sent to the psychiatry ward instead. He also treated a patient who, like Miss A in Belgium, was admitted for sudden psychosis and was thought to be schizophrenic. But his GI symptoms gave away the true cause of his illness – Whipple's disease again. Like the autistic children Dr MacFabe had cared for as a teenager, this young man was extremely perseverative, calling 'Dr MacFabe! Dr MacFabe! Dr MacFabe!' all day long. Within a week, antibiotics had cured him and he was back to his old self. Afterwards, he told MacFabe that he seemed like a character in a dream who had come to life.

These experiences firmly connected the gut and the brain in MacFabe's mind. The idea that chemicals produced by a single microbe could induce madness in a patient captivated him. When he heard about Sydney Finegold's discovery that autistic children had improved whilst on antibiotics, just as his Whipple's disease patient had done, MacFabe began to connect the dots. At the time, he was working on how the brain was injured during strokes, and he had been researching the effects of a molecule called propionate.

This molecule is one of a group of important chemicals produced by the gut microbiota when they break down the non-digestible remains of our meals. Known as short-chain fatty acids, or SCFAs, they include three major compounds: acetate, butyrate and propionate. Each has many roles to play, and all are fundamental to our health and happiness. What struck MacFabe though, was that although propionate was an important compound in the body, it was also used as a preservative in bread products – the very foods many autistic children crave. To top it all off, clostridia species are known to produce propionate. In itself, propionate is not 'bad', but MacFabe began to wonder whether autistic children were getting an overdose.

Was the altered microbiota in autism producing excess propionate? And could that propionate affect behaviour? MacFabe embarked on a series of experiments to find out. Via a tiny cannula positioned in the spinal column of live rats, he injected minute quantities of propionate into the fluid that bathed the brain. Within a couple of minutes, the rats started behaving strangely; spinning on the spot, becoming fixated on a single object, and dashing around. When two rats were placed together and given the propionate, they did not stop to sniff one another and interact as normal, but instead ran rings around the enclosure, ignoring their cage-mate.

There was nothing subtle about the response, and the similarity to autistic behaviours is striking – you can watch the videos online. The preference for objects over people, repetitive actions, tics, hyperactivity – all hallmarks of autism – were evident whilst propionate was acting on the brain. Within half an hour the effect had worn off, and the rats returned to their normal behaviour. For those injected with saline as a placebo, there was no change in behaviour whatsoever. Even injecting propionate under the skin, or feeding it to the rats, produced the same effect.

The brains of these rats had been hijacked by this tiny molecule, forcing the animals to behave abnormally. Was the propionate causing the same kind of damage to the rat brains as autism caused in human brains? By comparing the rat brains with the brains of autistic patients who had died and been autopsied, MacFabe and his team were fascinated to learn that both were stuffed with immune cells. Again, here was inflammation, just as in schizophrenia and ADHD.

Some inflammation in the brain is normal, as unnecessary synapses are gobbled up by the same immune cells that gobble up pathogens. Learning is a careful balance between remembering and forgetting. Making connections and seeing patterns is a hallmark of intelligence, but taken too far both processes will make a person unwell. When MacFabe put propionate-treated rats in a maze, he found they were able to learn the route through without a problem. But they then could not 'unlearn' it – if the route was altered, they persisted with their memory of the initial route, butting up against the newly added walls.

This is reminiscent of some autistic people's memories and love of routine. Flo and Kay Lyman are famous for being the world's only female autistic savant twins, and have been featured in several television documentaries. Despite struggling with social interaction and being unable to care for themselves, they possess extraordinary memories. Both women are able to instantly recall, for any given date, the weather, what they ate, and what their favourite TV show host wore on television. They know the name and artist of every song that has graced the charts, as well as its release date. These memories, once formed, have become permanently fixed, as if the synapses that hold them can never be discarded. Other synapses, meanwhile, such as those holding the rules for cooking a meal, do not survive.

Leo Kanner, in describing autism, noticed the same phenomenon. The children he studied seemed unable to adapt a definition once they'd learnt it. Most disturbingly, many of the children referred to themselves as 'you'. In saying 'Do *you* want to go and play?' and 'Would *you* like some breakfast?', parents inadvertently taught their child that its name was 'you'. The memory was inflexible. One child in Kanner's study even called her parents 'I', and herself 'you'.

Derrick MacFabe found that propionate-treated rats, who could not forget the original route through the maze, had an increase in compounds in the brain involved with memory formation. Problematic though this might seem, MacFabe believes there is an evolutionary purpose to it. If bacteria can release a compound that makes the brain remember, they can ensure that their host – the human body that carries them – will remember where to find the food that allows them to reproduce. In autism, he wonders, 'could over-activation of this pathway lead to "impaired forgetting", obsessional behaviours, food interests, and enhanced restrictive memories?' Indeed, the microbiota seem to be crucial for normal memory formation. Germ-free mice placed in a maze have trouble finding their way at all, because of deficits in working memory – their ability to hold information about routes they have already tried in mind as they move about. If MacFabe is right, simple changes in the composition of the microbiota could be flooding the body with propionate, changing the brain's ability to form and break synapses as a child develops.

So how do propionate and other compounds that could cause damage in large quantities travel from the gut to the brain? Just a couple of hours down the road from Derrick MacFabe is another scientist working on this question. Dr Emma Allen-Vercoe is a British microbiologist, working at the University of Guelph, who was introduced by Sydney Finegold over lunch to the idea that autism's cause lay in the gut. Like MacFabe, Allen-Vercoe suspects the combination of microbes in a child's gut is capable of producing compounds that tinker with brain function, the immune system and the child's human genes.

Rather than looking for a single species that is solely responsible, she is taking a holistic approach, viewing the gut microbiota as an ecosystem, like a rainforest. Taking any given species out of the rainforest and studying its behaviour while it's alone in a cage will not give away much about its true nature. This is true, too, of microbes, which are influenced by the presence of other microbes and the compounds they produce. So rather than study each species individually, Allen-Vercoe has re-created the microbiota's home, complete with all its inhabitants, outside the gut. A bubbling, stinky mass of tubes and bottles, this microbial home-from-home is affectionately known as Robogut.

By using Robogut, Allen-Vercoe has gone full circle, from the days when culturing bacteria in the lab was the only way to study them, through the DNA-sequencing revolution, and back to culturing. She disagrees that gut microbes are impossible to culture: 'That's absolute nonsense. You need a great deal of equipment, you need a good deal of patience, and you need a very good eye. We now have banks and banks of these "un-culturable" species in the freezer.'

Allen-Vercoe suspects the altered microbiota in the autistic gut is damaging the cells that line the colon, but instead of asking which *bacteria* are culpable, she is asking which of the *chemicals* the microbiota produce are responsible. The microbiota from a severely autistic child are taken, in the form of faeces, and given a new home within Robogut. In a rather crude semblance of our own inner tube, Robogut has a tube through which it is fed, and another through which it releases noxious gases. A third tube allows some of the liquid in which the microbes live to be filtered off. It is this 'liquid gold' that holds the chemicals, known as metabolites, that the microbiota have produced.

The hope is that by experimenting with the effect of the liquid gold on the cells of the gut in a Petri dish, Allen-Vercoe's research team will learn which metabolites are causing damage in the brains of autistic children, and exactly what they do. She has an enthusiastic graduate student, Erin, who is working on the idea. This is a girl with a vested interest in solving the puzzle of autism, because someone close to her is affected by it: her brother, Andrew Bolte.

In 1998, Ellen Bolte produced her first-ever scientific paper. Entitled 'Autism and *Clostridium tetani*', it was published in the journal *Medical Hypotheses*. It expounds her theory that autism results from the invasion of *C. tetani* in a child's gut after the razing of the normal, protective gut microbiota by antibiotics. Ellen's paper is a masterpiece of epidemiological and microbiological synthesis, drawing together evidence from dozens of studies to support each aspect of her hypothesis. As a piece of scientific rhetoric, Ellen's first contribution to her new field of research attests her past as a computer programmer, whose every thought must follow logically from the last. She can be credited with daring to open a Pandora's box of medical possibilities, not least the concept that there may be behavioural consequences stemming from alteration of the human body's microbes. Ellen's achievement is testament not only to her own intelligence and determination, but also to the power of a mother's need to protect her child.

But, as Derrick MacFabe, whose research has been influenced by Ellen's foresight, points out, 'Hypotheses, although essential, are not enough. They must be tested.'

Fortunately, Ellen Bolte has passed on both the legacy of her hypothesis and her innate sense of scientific rationale to Erin, her daughter. Erin has found her calling in trying to solve the mystery of the illness that changed the course of her younger brother Andrew's life, twenty-odd years ago. She has begun her scientific career in earnest, under the guidance of Dr Emma Allen-Vercoe at the University of Guelph in Ontario, Canada. Using Robogut, her aim is to take her mother's hypothesis, in its broadest sense, and test it.

Erin wants to understand exactly what took place in the gut of her brother and the eleven other young children whose autism improved during the eight-week antibiotic trial they took part in. She also wants to know why the parents of autistic children report an improvement in symptoms when they cut certain foods out of their child's diet. Robogut gives her the means to see exactly what changes in the autistic gut microbiota when she adds antibiotics, gluten and casein (wheat and milk proteins) to the mix. Given that autistics can improve on antibiotics, which metabolites are no longer being produced when Erin gives Robogut the same drugs? Given that autistics often get worse when they eat baked goods, which metabolites are produced in greater quantities when she feeds Robogut gluten?

Erin's experiments will lay a foundation not only for understanding the role of the microbiota in autism, but for discovering its contribution to many other neuropsychiatric conditions. Her mother, Ellen, went way beyond the call of duty, bringing her sleuth-like logic to an extremely complex illness, and opening up an extraordinary new avenue of investigation when no one wanted to listen. Now Erin has picked up the baton, and is applying her own considerable intelligence and determination to finding answers to the questions that ever more parents are having to ask. For Andrew, whose childhood developmental window has closed, life will probably always be lived within the confines of his autism. For Erin, though, as well as for Derrick MacFabe and Emma Allen-Vercoe, the hope is to prevent this insidious illness from affecting, as predicted, every family in America, and others around the world.

In health, we like to think we are the products of our genes and experiences. Most of us credit our virtues to the hurdles we have jumped, the pits we have climbed out of, and the triumphs we have fought for. We see our underlying personalities as fixed entities – 'I am just not a risk-taker', or 'I like things to be organised' – as if these are a result of something intrinsic to us. Our achievements are down to determination, and our relationships reflect the strength of our characters. Or so we like to think.

But what does it mean for free will and accomplishment, if we are not our own masters? What does it mean for human nature, and for our sense of self? The idea that *Toxoplasma*, or any other microbe inhabiting your body, might contribute to your feelings, decisions and actions, is quite bewildering. But if that's not mind-bending enough for you, consider this: microbes are transmissible. Just as a cold virus or a bacterial throat infection can be passed from one person to another, so can the microbiota. The idea that the make-up of your microbial community might be influenced by the people you meet and the places you go lends new meaning to the idea of cultural mind-expansion. At its simplest, sharing food and toilets with other people could provide opportunity for microbial exchange, for better or worse. Whether it might be possible to pick up microbes that encourage entrepreneurship at a business school, or a thrill-seeking love of motorbiking at a race track, is anyone's guess for now, but the idea of personality traits being passed from person to person truly is mind-expanding.

FOUR

The Selfish Microbe

Everyone wants to know how to improve their immune system. Type 'immune system' into Google and the first suggested phrase combines it optimistically with the word 'boost'. In a perfect world, we would achieve this boost with a sweet-tasting superfood, preferably some kind of berry from a secret location in the Andes. The exorbitant cost of such a berry could only mean one thing: it must work! For most people, boosting the immune system is all about avoiding the endless rounds of colds and the occasional flu caught from germ-riddled grab bars on buses and trains. But what really holds the key to having a healthy, active immune system?

It is an unfortunate consequence of our super-sociality that we are so prone to picking up bugs. A few unpleasant weeks of feeling not quite well enough to work, but not quite ill enough to spend the day on the sofa, are inevitable at least once annually. Of course, by biting the bullet and dragging ourselves into the office, snotting and snuffling our way through the day, we are doing exactly what the pesky microbe 'wants' us to do – continuing to be social, and therefore spreading it far and wide. Not-quite-ill-enough-to-stay-at-home means the pathogen (that's a disease-causing microbe) has achieved the perfect balance of virulence and innocuousness. They are at once virulent enough to get themselves passed on – coughs and sneezes spread diseases! – and innocuous enough to ensure you don't die before encountering other people – all potential hosts for the bug. One of the small mercies of really dreadful infectious diseases with mortality percentages up in the 90s, such as Ebola and anthrax, is that they are so very virulent and kill so very rapidly that they barely have the chance to infect anyone else. Ebola's fearsome efforts in sustaining the epidemic that began in West Africa in 2014 are likely due to it becoming *less* virulent, dropping to a death rate of around 50–70 per cent of the people it infects. This drop in virulence, and mortality rate, means victims live just that little bit longer, giving the virus a better chance of switching to a new host and perpetuating its spread.

Many wild animals, on the other hand, don't tend to suffer these irksome illnesses, not because they have superior immune systems but because it takes contact between fresh, susceptible individuals to sustain an outbreak of disease. Some lonesome mountain goats in the French Alps don't ever meet their cousins in the Pyrenees, so infections are rare. Likewise, for species that prefer a largely solitary existence – leopards, say – infectious diseases just can't get a foothold in their population.

Couple super-sociality with wanderlust and the pathogens rejoice. Constant contact, and an unlimited supply of new blood. It's no accident that alongside humans, bats are one of the most potent carriers of disease worldwide (probably including Ebola). Like us, many bat species live in vast colonies of thousands or millions of individuals, crammed into densely packed spaces. This gives pathogens ample opportunity to take up residence, spreading in Mexican waves through the bats, then mutating and doing another lap months or years later. What's more, bats can fly. As individuals from different roosts gather at feeding grounds, so too do their microbes, allowing them to bridge the gap between otherwise isolated populations. Human beings share these features – super-sociality and hyper-mobility – with bats more closely than feels comfortable. We crowd together in cities and jet around the world, sharing and spreading microbes, both pathogenic and harmless, as we go.

The reality for most people is that, far from having an under-active immune system, they have an overactive one. It might seem normal to suffer an attack of hay fever each spring, and to sneeze every time you pick up the cat, but it is not. Normal, perhaps, in the sense that vast numbers of people in the developed world suffer with allergies. But take a step back and consider what kind of evolutionary process would leave 10 per cent of children gasping for breath once in a while, as asthma does. How could it ever be beneficial for 40 per cent of children, and 30 per cent of adults, to be intolerant to something as ubiquitous as pollen, as is the case in hay fever? People with allergies rarely see themselves as having an immune dysfunction, but that is exactly what it is. What they need, though, is not a 'boost', but quite the opposite. Allergies are the result of an immune system that's over-eager, acting to destroy substances that pose no real threat to the body. Indeed, treating them often involves calming the immune system down again, through steroids or antihistamines.

In the most developed countries, allergies are already entrenched. By the 1990s, their rise had slowed enough to reach a plateau in the proportion of the population affected. This levelling off could simply indicate that all those genetically susceptible to allergy are now affected by it, rather than that the underlying cause has stabilised. But far from the madding crowd, in truly rural, pre-industrial, non-Westernised parts of the world, allergies just aren't the problem they are elsewhere. For everywhere in between the developed world and increasingly rare pockets of tribal culture, the merciless rise persists, sweeping more and more people into this state of unnatural immune overreaction as the generations go by. Since the rise began in the West around the 1950s, the question has always been: What is the underlying cause?

For most of the last century, the traditional theory was that allergies are triggered in children when they suffer from infections. In 1989 a British doctor named David Strachan challenged that theory. He did so, in a brief and straightforward paper, by suggesting the exact opposite: that allergies were the result of *too few* infections. Strachan had studied a national database containing health and social information about a group of over 17,000 British children born in a single week in March 1958, who were followed until they reached the age of twenty-three. Of all the data collected on these children – social class, wealth, location and so on – two things stood out as being connected to their chance of suffering from hay fever. The first was the number of brothers and sisters a child had – an only child was far more likely to have hay fever than a child with three or four siblings. The second was the child's position in the family – children with older siblings were less likely to have hay fever than those with younger siblings.

As anyone who has children knows, toddlerhood can bring a constant stream of sniffles. Young children are a hotbed for bacteria and viruses, as their immune systems are naive to the onslaught of pathogens facing humans every day. Toddlers' habit of putting everything within reach into their mouths means that their microbes, both good and bad, are spread liberally wherever they go. The more children, the more microbes are left in snail trails of snot and saliva. David Strachan's suggestion was that children in bigger families were benefiting from the extra infections that their siblings – particularly the older ones – brought home. Somehow, he thought, these infections in the early years of a child's life carried with them protection from hay fever and other allergies.

Quickly dubbed the 'hygiene hypothesis', Strachan's idea was supported by the fact that the rise in allergies was matched by improvements in hygiene standards over time. A weekly wash before church in a lukewarm bath became a daily shower in steaming hot water. Food was refrigerated or frozen rather than pickled and fermented. Family sizes were shrinking, and life was becoming more and more urban and refined. The hygiene hypothesis just made sense, especially as in developing countries with high rates of infectious disease, allergies were still rare. It seemed that in Europe and North America, people were simply too clean for their own good, and their immune systems were champing at the bit, desperate to attack even harmless particles like pollen.

Although the hygiene hypothesis represented a new paradigm for immunologists, it quickly gained scientific favour. It has an intuitive appeal – the immune cells are easily personified as aggressive hunters, eager to seek and destroy. They are, we imagine, restless when undertasked. And so, as the serious infectious diseases are shot down by vaccinations, and cleanliness takes care of more benign germs, immune cells are left all riled up with nothing to kill. This anthropomorphism makes it all too easy to accept the suggested outcome: undertasked immune cells turn their attention to harmless particles and continue the great fight.

The idea has been extended from bacterial and viral infections to parasites, particularly worms, of the tape, pin and hook variety. As with microscopic pathogens, your chances of taking on board a worm have dwindled to nearly nothing in the developed world. It has left a suspicion amongst both scientists and the public that worms had been keeping the immune system occupied, and their absence has now left it overstaffed and underworked.

The connection that David Strachan had uncovered between family sizes and allergies bore up in dozens of other studies. A neat theory emerged of exactly how that idea might work. Imagine for a moment that the immune system has two divisions: the army and the navy. Now, please forgive me for the following oversimplification of the armed forces. Let's assume that the army deals with threats on land, and the navy with threats at sea. A decrease in threats at sea means that many of those who would have joined the navy instead are recruited into the army. But no additional threats on land leaves the army overmanned.

So it goes in the body: a division of immune cells known as T helper 1 cells (T_h1) usually respond to threats from bacteria and viruses. A second division – T helper 2 cells (T_h2) – usually respond to parasites, including worms. If the infectious disease burden from bacteria and viruses drops because of improved hygiene, the T_h1 division shrinks. The T_h2 division takes up the slack, but the extra T_h2 cells find themselves with too little to do. So, as well as watching for worms, they begin to attack harmless particles, pollen and dander included. Simple and elegant, but was it realistic?

The next challenge for Strachan was to look for a clear link, not just between family size and allergies, but between *infections* and allergies, to back up his hypothesis. Some data appeared to offer support for the idea: people infected with hepatitis A or measles didn't have allergies as often as the uninfected. Perplexingly though, evidence for this connection was hard to find for the majority of common infections. Indeed, Strachan himself found that babies who had suffered an infection in the first month of their lives were no more likely to develop allergies than those who hadn't. Even on those occasions where more infections did mean fewer allergies, there was often a better explanation for the link.

Unfortunately, like my army and navy analogy, the simplistic division of the immune system's anti-pathogen chores between T_h1 cells and T_h2 cells does not bear out. No pathogens are fought solely by T_h1 cells or just by T_h2 cells – all provoke a few of each type. What's more, if excess T_h2 cells were to blame for rising rates of allergies, then childhood diabetes and multiple sclerosis would not also be on the rise. Both of these are autoimmune diseases, in which the body attacks its own cells, and both involve an excess not of the T_h2 division, but of the T_h1 division.

The most contradictory element to the hygiene hypothesis, though, is that in the absence of germs and worms, there remains an apparently legitimate target for immune cells to pursue in preference to pollen and dander. Though true pathogens are now relatively rare visitors to Western humans, their bodies contain a microbial invasion the size of which should intimidate even the most voracious of immune systems. Weighing a couple of pounds in total, with their control centre in the colon, this invasive group of 'germs' – the microbiota – live alongside the greatest concentration of immune cells in the human body. And yet they survive, unharmed. If the immune system were all hyped up with no enemies to attack, wouldn't they relish the chance to take on these intruders?

It all comes down to the question of how the immune system knows what to attack. It may strike you as obvious: it should attack anything that is not part of the body. Anything non-human in humans, non-cat in cats, non-rat in rats. *Non-self*, in other words. All it would need to do is recognise what is human (or cat or rat) – *self* – and tolerate it, and recognise what is non-human – *non-self* – and destroy it. It is this *self/non-self* dogma that has framed immunology for over a century.

But now consider what would happen if the immune system actually applied this classification system; if it attacked anything it encountered that was *non-self*. Food molecules? Pollen? Dust? Another person's saliva, even? Reacting to these is not helpful and is a complete waste of energy – they are, in themselves, harmless. Just because something is *non-self* does not imply that it is dangerous and must be attacked. Some things are better left alone.

On the other hand, imagine what would happen if the immune system destroyed nothing that was *self*. It's a bit more obscure, but no less important, for there are more things than you might think that are *self* but are nonetheless unwanted. For starters, if it weren't for the immune system's removal of *self*, we would all have webbed fingers and toes. By around nine weeks into a pregnancy, a human foetus has just started to resemble a human, but is around the size of a grape. At this point, the cells between the fingers and toes 'commit suicide' – undergoing programmed cell death to allow the fingers to separate. The clean-up operation is run by phagocytes – a group of immune cells that engulf and disassemble the discarded webbing.

Likewise, the synapses of the brain. Achieving that perfect balance between remembering and forgetting means destroying the junctions between neurons that are no longer useful – a task carried out by a specialised form of phagocyte. The same goes for cells in danger of becoming cancerous. More often than you might care to know (we're probably talking dozens of times a day), mistakes made when DNA is copied threaten to give a cell the key to immortality – cancer. It is the cells of the immune system, patrolling the body looking for signs of these errors, that prevent this from happening nearly all of the time.

Tolerating *some* non-self substances and attacking *some* self molecules is just as important as destroying non-self pathogens.

The microbiota is quite clearly *non-self*. They are entire organisms, belonging not only to different species from us, but to several different kingdoms of life. What's more, they are extremely similar to the very creatures that cause the immune system so much trouble – the pathogenic varieties of bacteria, viruses and fungi. Members of the microbiota even have the same type of give-away molecules coating their surfaces that the immune system uses to detect pathogens. But something about these microbes tells the immune system not to attack.

David Strachan's original hygiene hypothesis was an excellent one, but it now faces an overhaul. He suggested that more infections in childhood meant a lower chance of allergies. The trouble is, the evidence doesn't support the idea, and the mechanisms don't quite work. But in some sense the rethink the hygiene hypothesis is undergoing is a subtle one. Although it does not cause disease, the microbiota is, in some sense, a vast infection. These microbes are intruders, but they have been intruding for such a long time, and they bring such great benefit, that the immune system has learnt to accommodate them. So how have our resident microbes evaded destruction? And what happens to our finely tuned immune systems if they get out of balance?

There came a moment in my discovery of the microbes that inhabit the human body when I stopped seeing myself as an individual, and began to consider myself instead as a vessel for my microbiota. Now, I see us – myself and my microbes – as a team. But, as in any relationship, I will only get what I give. I am their provider and their protector, and in return they sustain and nourish me. I find myself thinking about my meal choices in terms of what my microbes would be grateful for, and my mental and physical health as markers of my worthiness as a host to them. They are my own personal colony, and their preservation is worth as much to me as the well-being of the cells of my own body.

Despite my uniquely human ability to consciously comprehend this partnership, it is not a uniquely human alliance. As I will explain in chapter 6, your own colony begins with a Noah's Ark of species gifted to you by your mother at birth. Your mother's first microbes, of course, came from your grandmother, and hers from your great-grandmother, and so on *ad infinitum*. Somewhere in the realm of 8,000 'greats' ago, that microbial gift was being transferred from mother to child in our pre-*Homo sapiens* ancestors. This transfer continues back into our evolutionary history, beyond humanity, beyond primates, beyond mammals, even, at least to the dawn of the animal kingdom.

It is a favourite trick of biology teachers to ask their students to hold their arms out wide, and to trace the biography of the planet across their span. At the far tip of the right hand's middle finger lies the formation of the Earth, 4.6 billion years ago. The tip of the left hand represents the present day. It took until the elbow of the right arm for Earth's rocks to cool, and life to begin in the form of bacteria. From these simple beginnings, it took up to a little shy of the left wrist – another 3 billion years – before the most basic of animals evolved. Mammals, in all their furry, thoughtful glory, came about just onto the left middle finger, and humans like us stepped up in the final hair's breadth of this finger's nail. It's been said before: one swipe of the nail file, and all traces of our existence would vanish.

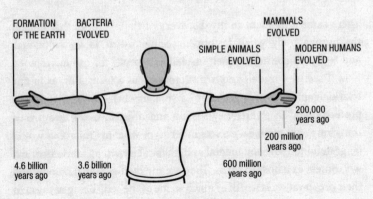

A history of planet Earth.

Animals, then, have never known life without bacteria. So inter-twined are their existences, that smuggled inside (nearly) every cell of every animal are the ghosts of the very simplest of bacteria. Engulfed by a larger cell, these bacteria are worth playing host to, as each one specialises in making energy from molecules of food. They are your mitochondria – the power stations of your cells – converting food to energy by cellular respiration. Just as fundamental to multicellular life now as in the early days of Kingdom Animalia, these ex-bacteria have become so ingrained that we no longer consider them to be microbes in their own right. Mitochondria are an evolutionary signature of the earliest alliances between two organisms. Ever since, the smallest of organisms have been teaming up with larger ones.

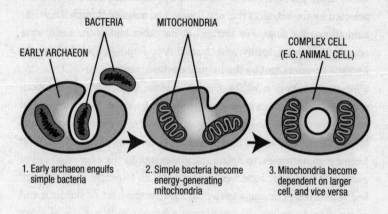

BACTERIA   MITOCHONDRIA

EARLY ARCHAEON

COMPLEX CELL
(E.G. ANIMAL CELL)

1. Early archaeon engulfs simple bacteria

2. Simple bacteria become energy-generating mitochondria

3. Mitochondria become dependent on larger cell, and vice versa

The evolution of mitochondria.

The pattern of these alliances can be seen in the tree of life. Draw-ing an evolutionary tree of the relationships between species of mammal, and another tree of the bacteria that inhabit those mam-mals, reveals that the two groups have travelled together through evolutionary time. The trees mirror one another – where one mammal species splits into two, the microbes that it hosts also divide, evolving separately with their new mammalian host from that point on. This close-knit relationship between a host and its microbiota has led to a revolutionary new idea that gets to the core of the workings of evolution by natural selection.

I'll begin how Darwin began in *On the Origin of Species* – not with natural selection, but with artificial selection. Darwin wrote about breeding pigeons, for that was the favourite hobby of a certain class of gentleman in his day. I shall illustrate my point with dogs instead. A great Dane and a fox terrier are both descended from wolves, yet neither looks anything like its ancestors. Great Danes were bred to hunt deer, boar and even bears in the forests of Germany. Each generation, their breeders choose dogs with size, speed and power to become the parents of the next generation. Steadily, these features become more pronounced, as the most appropriate characteristics are selected by the breeders (i.e. artificially, not naturally) from the variation among the dogs. Fox terriers, at the other end of the scale, were selected for speed, agility, and the ability to fit down a fox hole, again by their breeders, not by the natural environment.

Natural selection works much the same way, except rather than a breeder choosing which characteristics to select for, the natural environment does the job. A cheetah must have powerful enough legs to run after prey; a big enough heart and lungs to outlast its quarry; keen enough eyesight to notice the youngster on the edge of a herd of gazelle. A frog must have webbed enough feet to propel itself through the water, tough enough spawn to survive the heat of the sun, and camouflaged enough skin to escape a heron's notice. Each of these traits is selected for: by the climate, the habitat, the competitors, the prey, or the predators of the organism.

What evolutionary biologists wrangle with is what *exactly* is being selected for. It might seem clear: the *individual* with the powerful muscles or the webbed feet is the one who will live and breed. It is that individual that has been selected by its circumstances to reproduce. But then why would lionesses help their cubs? Why would worker bees help their queen? Why would young moorhens help their parents? Even more perplexingly, why would one vampire bat regurgitate a blood meal for another that failed to feed, even though the two are unrelated? If all that mattered to an individual was its own reproductive success, why help others? It is these questions that get biologists talking about selection above the level of the individual. If cooperation helps members of a group, especially where the members are unrelated, to reproduce, it is not just individuals that are selected by their environments, but entire groups.

Richard Dawkins would remind us that both individual and group selection are missing the point. In his 1976 book *The Selfish Gene*, he argued the perspective of several prominent evolutionary biologists – that it is *genes* that natural selection ultimately chooses between. Bodies, he says, are mere vehicles for genes, enabling them to live an immortal existence. They, like individuals, have variation, can be replicated, and are passed between generations. The point is that it is genes that determine an individual's likelihood of reproduction, and therefore that *they*, not the individual, are selected for or against. Of course, a gene cannot act alone. Even the most life-giving, baby-producing gene that mutation could bestow upon an individual is limited by the virtues of its neighbouring genes. Which brings us back to the beginning – the individual – but perhaps with a greater appreciation of the complexity of natural selection's predicament.

The paired evolution of host species and their microbiotas adds a further layer to this already tangled web of evolutionary favours. Let's take a herbivore – a bison – to imagine natural selection's task in the absence of the microbiota. Our bison would be large enough to fend off wolves, hairy enough to keep warm through the winter, and strong enough to travel long distances in pursuit of good grazing. But these, and many other life-giving and baby-producing characteristics of the bison, are futile without a decent set of gut microbes. Alone, the bison cannot digest grass. With a belly full of untouched grass, it cannot absorb food molecules, and it cannot make energy. Without energy, it cannot grow, move, reproduce or stay alive. It is useless.

Bison and their microbiotas have evolved together. They have been selected for together. The bison must be large enough, and hairy enough, and strong enough, but it must also be good enough at digestion to earn natural selection's favour. It must be *microbial* enough. The combination of the host (the bison, fish, insect or human) and its microbes is known as the 'holobiont'. It is this co-dependence, this evolutionary inevitability, that led Eugene and Ilana Rosenberg of Tel Aviv University in Israel to propose another level on which natural selection can act. Not only are individuals chosen for their reproductive merits, and groups for theirs, so too are holobionts. As no animal can ever live independently of its microbiota, and no microbiota without its host, selecting one without the other is impossible. Natural selection, then, acts on both, choosing, as it does with individuals, those combinations of vehicle and passengers that are strong enough, suitable enough, fit enough, to survive and reproduce.

Ultimately, as Dawkins makes clear, it is the genes that selection acts upon, whether animalian or microbial. Accordingly, the Rosenbergs' idea is known as 'hologenome selection': selection on the host genome and the microbiome combined.

The point is, the human immune system did not evolve in isolation. It was never a sterile set of nodes and tubes and roving cells, waiting for an assault by an unknown enemy. Rather, it has 'grown up' with microbes of all kinds – both those that make us distinctly ill, and those that keep us well. Because of this millennia-long association, the immune system expects microbes to be present. When they are not, the balance is off. It is as if you had learnt to drive with the handbrake permanently on. You have mastered exactly how much pressure to apply to the gas pedal to overcome the brakes, but suddenly, after decades of driving around safely with it on, the handbrake is released, and your driving becomes wild and erratic as you struggle to keep up.

Of course, even the most unhealthy among us are not entirely without a microbiota, so we cannot know for sure what havoc the immune system might wreak in their absence. In the history of humanity, just one person has lived out their life without a microbiota, or just about. To do so, this young boy spent his life in a bubble in a Houston hospital. Known by the media as the 'Bubble Boy', David Vetter suffered from Severe Combined Immunodeficiency (SCID), which meant he was utterly unable to defend himself against pathogens. David's parents had lost their first son to the genetic disease, but doctors hoped that a cure was just around the corner, so David's mother went ahead with another pregnancy.

David was born in 1971 by Caesarean section into a sterile plastic bubble. He was handled through plastic gloves and fed sterilised infant formula. He never knew the scent of his mother's skin, or the touch of his father's hand. He never played with another child without plastic sheeting preventing the sharing of toys and laughter. To get David out of his bubble, a bone-marrow transplant from David's sister was needed. The hope was that it might kick-start his immune system and free him of the disease. But his sister was not a transplant match. David had no choice but to remain inside his bubble for the rest of his life.

Despite his devastating illness, David lived out his highly protected life in comparative health, and was not unwell once until his death at age twelve. Intriguingly, he had a very poor spatial memory, but was hypersensitive to the passing of time, perhaps because his brain had no need to find its way around in space, only time. Many studies of David's physical and mental health were carried out during his isolation, but the known benefits of a microbiota were limited to its role in synthesising vitamins, and the impact of being germ-free on David was not investigated. Eventually, in the absence of a matching donor, it was decided that David should have a bone-marrow transplant from his sister, despite the risks. A month after the surgery, David died from lymphoma – a cancer of the immune system. The cancer was provoked by the Epstein–Barr virus, which had stowed away in his sister's bone marrow, and unbeknown to the doctors, had been transferred to David.

Despite their best efforts to keep David germ-free, from birth onward his gut had been colonised by more and more species of bacteria. His doctors knew this, as they would periodically culture the bugs in David's faeces, but the simple colony he had acquired did not seem to be doing him any harm. Had David been truly germ-free, the coroner at his autopsy might have discovered that David's digestive system was drastically out of proportion. The first tennis-ball-like section of the large intestine – the caecum – to which the appendix is attached, might have been more like a football than a tennis ball. The folded surface of the small intestine would probably have had a much smaller surface area than normal, and fewer blood vessels supplying it. As it was, David's digestive system was as normal as any other child's.

These gastrointestinal differences are typical of germ-free animals, though it's not clear why. One researcher told me that the first time she dissected a germ-free mouse, she was horrified by the size of the caecum, which took up most of the space in the abdomen. She later learnt that all germ-free mice have super-sized caeca. Their immune systems, too, are distinctly different from those of normal animals, almost as if they have never matured after birth. The small intestine of normally colonised mammals, including humans, is usually dotted with clusters of cells that act as border-patrol stations, called Peyer's patches. In each patch is a row of miniature assessment centres, into which passing non-self particles are dragged and 'interviewed' by immune cells about their intents and purposes. Those which appear suspicious spark a hunt for more within the gut, and occasionally the rest of the body. In germ-free animals, though, these border stations are few and far between, and the guards they contain are poorly trained and slow to alert their colleagues to a breach in the border. With an immune system like this, a germ-free animal's prospects outside the safety of its isolation bubble are not good. In fact, when released from their cocoons, germ-free animals quickly succumb to infection, and die.

Clearly, the microbiota alters the immune system's development. This has a powerful effect on its ability to fight disease. When infected with *Shigella* bacteria – which causes severe diarrhoea in humans – normal guinea pigs are unaffected, but germ-free guinea pigs invariably die. Even adding just one species of the normal microbiota to the guinea pigs will protect it from the fatal effects of the *Shigella*. This effect is not just seen in animals with no germs at all, though. Giving animals antibiotics that shift the normal balance of microbiota also leads to more infections. Mice treated with antibiotics, for instance, cannot fight off the flu virus if it is inserted in their noses. They get sick, whilst those not on antibiotics do not. Immune cells and antibodies – which tag pathogens for destruction – are just not produced in large enough numbers to prevent the infection spreading into the lungs.

It seems paradoxical – surely antibiotics are there to *treat* infections, not cause them. But although a course of antibiotics might cure one infection, they may also leave us open to others. The obvious explanation is that without protective microbes, the body is left exposed to attack by pathogenic ones. But antibiotics rarely decrease the total number of protective microbes present; instead, they shift the composition of the different species. It seems the real change is in how the immune system behaves, according to which members of the microbiota are present.

The first line of defence in the gut is a thick layer of mucus. Close to the cells of the gut lining the mucus is free of any microbes, but in the outer layer, many members of the microbiota make their homes. Treatment with, for example, the antibiotic metronidazole kills just the anaerobic bacteria – those that live without oxygen. This shift in the composition of the microbiota changes the fundamental behaviour of the immune system. It interferes directly with our human genes, dialling down production from the genes that churn out mucin proteins for the protective mucus layer. With a thinner layer left behind, microbes of all kinds have easier access to the gut lining. If they, or their chemical components, get through to the bloodstream on the other side, the immune system is mobilised.

You'd be forgiven for wondering, if antibiotics changed the functioning of the immune system, wouldn't we be getting sicker as a result of taking them? In a study of 85,000 patients, those taking long-term antibiotics for the treatment of acne were more than twice as likely to suffer colds and other upper respiratory tract infections as acne patients who weren't on antibiotics. Another study of college students found that being on antibiotics quadrupled the risk of getting a cold.

But what about the effect of antibiotics on allergies? In 2013, a group of scientists at the University of Bristol in the UK asked just that question. They used a massive research project known as Children of the 90s. This project collected a huge amount of health and social information about the children born to 14,000 women who were pregnant in the early Nineties. The data included information about antibiotic use when the children were infants. It turned out that those who had been given antibiotics before the age of two – a startling 74 per cent of them – were on average nearly twice as likely to have developed asthma by the time they were eight. The more courses of antibiotics the children received, the more likely they were to develop asthma, eczema and hay fever.

But, as the saying goes, correlation does not always mean causation. The lead researcher on the antibiotics study had discovered four years earlier that the more television children watched, the more likely they were to develop asthma. Of course, despite similar results as in the antibiotics study, no one really believed that the act of watching television could bring about immune dysfunction in the lungs. In fact, the number of hours in front of a television was being used as a proxy for the amount of exercise children were getting. The point still stands, though – how can we be sure that the amount of antibiotics a child received isn't also a proxy for something? For example, it may relate to how fussy the parents were, or, perhaps more pertinently, the possibility that the antibiotics were given to treat early symptoms of asthma. But the researchers accounted for this by calculating the likelihoods again, this time excluding any child who had suffered from wheeze before the age of eighteen months. The link remained strong.

Of course, taking antibiotics is all about ridding the body of infection, so the hygiene hypothesis stands up in light of the link between them and allergies. But the paradox remains: why would the immune system attack harmless allergens in preference to the apparently more alarming threat from the body's microbes? And if the rise in allergies were connected to the drop in infections, why wouldn't those of us who had had fewer infections be the ones who had more allergies?

Professor Agnes Wold of the University of Gothenburg in Sweden was the first to provide an alternative to the hygiene hypothesis, in 1998. Research into the importance of the microbiota was just beginning to pick up steam, and the lack of correlation between infections and allergy had started to challenge Strachan's idea. Alongside the correlation between antibiotic use and allergy was a more direct connection. Wold's colleague Ingegerd Adlerberth had previously compared the microbes of babies born in hospitals in both Sweden and Pakistan.

The babies born in Sweden, where allergy rates are high, had a much lower diversity of bacteria than the babies born in Pakistan, especially of a group called the enterobacteria. For sure, Sweden's level of hygiene far exceeded that in Pakistan, but these babies weren't getting sick – they weren't suffering from infections. They were simply being colonised by more microbes, especially from a group of bacteria that are found in the adult gut, and therefore the mother's faeces, as well as the general environment. Midwifery protocols in Sweden, including in some cases the practice of cleaning women's genitals before they give birth, may have been utterly altering the microbes that set up camp in the empty guts of newborns.

Wold felt that it was this change in the composition of the microbiota, rather than exposure to outright infections, that was driving the rise in allergies. She organised a large study of the babies in Sweden, Britain and Italy, and tracked the changes in their microbiotas over time. As expected, these ultra-hygienic babies were colonised by fewer species – again, especially of the enterobacteria. In their place were more members of a group of bacteria that are typically found on the skin, not in the gut – the staphylococci. No single species or group of microbes could be linked to the development of allergies later in life, but the overall *diversity* of microbes in the babies' guts could be. The babies who went on to develop allergies had far lower diversity in their guts than the babies who remained healthy.

Wold's reframing of Strachan's original hygiene hypothesis has gained ground amongst immunologists and microbiologists. Two decades of research into the body's microbiota have added another layer of complexity to our understanding of what triggers the immune system to develop in a healthy way. It is not simply infectious diseases that a super-clean environment puts a stop to, but normal colonisation by the microbes that are sometimes referred to as 'Old Friends'. These Old Friends have evolved with us every step of the way, deeply engaged in conversation with the immune system as they have gone along. The hygiene hypothesis has mutated into the Old Friends hypothesis – a new take on an old idea. Exactly what the microbiota is telling the human body, or the body of any animal for that matter, is the next question. How does the body know which microbes to trust and which are true impostors?

The immune system operates a team of different cell types, each with a specific role in detecting and destroying threats, much like the armed forces. Whereas *macrophages* are foot soldiers which gobble up threatening bacteria, and *memory B cells* are snipers, trained to attack a specific target, *T helper cells* (such as T_h1 and T_h2) are communications officers, alerting other troops to an invasion. Triggering all of these responses begins with antigens – small molecules on the surface of a pathogen that identify it as an enemy. These antigens act like little red flags that the immune system automatically recognises as danger signals, whether it has encountered that particular pathogen before or not. Since all pathogens carry these flags, they will always be detected once they enter the body.

Back when the concept of self and non-self ruled the thinking of immunologists, it was thought that pathogens gave away their presence through the antigens that they wore on their cell surfaces. But what researchers didn't acknowledge at that time was that beneficial microbes also carried these flags, and they sent just the same message to the immune system as those on the pathogens. The flags merely identified them as microbes, and told the body nothing about whether they were friend or foe. The microbiota was not simply being ignored by an enlightened immune system. Instead, our beneficial microbes must have found a way to persuade the immune system into letting them stay.

You might think that all immune cells would be out to destroy targets and hunt for threats. But, as with every system in the body, there must be a balance – dials that turn up and dials that turn down. It is no different in the immune system: pro-inflammatory (attack) messages must balance anti-inflammatory (stand-down) messages. The anti-inflammatory side of the equation is brought about by a newly recognised immune cell, the regulatory T cell, which acts as a brigadier, coordinating the overall immune response. Known as T_{regs} (pronounced T-regs), these cells exert a calming influence over the aggressive, bloodthirsty members of the immune system's ranks. The more T_{regs}, the less reactive the immune system, and the fewer, the more aggressively it responds.

Indeed, in children who have a genetic mutation that prevents them from producing T_{regs}, a catastrophic disease known as IPEX (immunodysregulation polyendocrinopathy enteropathy X-linked) syndrome results. The immune system is pushed off balance and it produces massive quantities of pro-inflammatory immune cells, causing the lymph nodes and spleen to swell with the excess. The aggressive cells fatally attack the body's organs, usually bringing on early diabetes, eczema, food allergies, inflammatory bowel disease and intractable diarrhoea during childhood. These multiple autoimmune and allergic disorders result in an early death as successive organs are destroyed.

The surprising new evidence though, is that the commander-in-chief commanding the anti-inflammatory T_{regs} is not a superior human cell type acting solely in the body's best interests. Rather, it is the microbiota that delivers the orders, using the T_{regs} as its pawns. By manipulating the numbers of this suppressive rank of cells, they ensure their own survival. For them, a calm, tolerant immune system means an easy life, without fear of attack or elimination. The idea that our microbiota has evolved to suppress our immune systems for its own benefit seems a little intimidating. A safety mechanism that is fundamental to our survival in the face of attack by our oldest adversaries is being tampered with at the highest level. But we needn't worry – our long co-evolutionary history with our microbiota has fine-tuned the immune balance, to the best advantage of both us and it.

More concerning is the loss of diversity that has hit the microbiotas of humans living a Westernised lifestyle. With lower diversity, what becomes of the T_{regs}? A group of scientists including Agnes Wold asked this question using the laboratory's old favourite – germ-free mice. They measured the effectiveness of T_{regs} from germ-free mice and compared them with normal mice. In the germ-free mice, a far greater ratio of T_{regs} to attacker cells was needed to suppress the aggressive immune response, showing that the T_{regs} produced in the absence of a microbiota were much less powerful than those in a normal mouse. In another experiment, simply adding a normal gut microbiota to a germ-free mouse increased the number of T_{regs} the mouse produced, and this calmed the immune system's aggressive behaviour.

So how do the members of the microbiota do it? Their cell surfaces are covered in red flags, just as are found on pathogenic microbes, yet they manage to pacify rather than irritate the immune system. To do so, it seems that each beneficial species may have its own password, known only to it and the immune system. Professor Sarkis Mazmanian, at the California Institute of Technology, discovered such a password, produced by a bacterium called *Bacteroides fragilis*. This species is one of the most numerous members of the microbiota, which often takes up residence in the gut immediately after birth. It produces a chemical called polysaccharide A, or PSA, which is released from the surface of *B. fragilis* in tiny capsules. The capsules are engulfed by immune cells in the large intestine, and the PSA inside triggers the activation of T_{regs}. The T_{regs} then send out calming chemical messages to other immune cells, preventing them from attacking the *B. fragilis*.

By using PSA as a password, *B. fragilis* converts the pro-inflammatory response of the immune system into an anti-inflammatory one. It's likely that PSA and other passwords produced by other early colonisers are important in calming the immune response and preventing allergies. Many varieties of these passwords have probably evolved to be produced by individual strains of microbes to gain acceptance as a member of the exclusive club within the human body. Just as in the fatal immune disease IPEX, animals with allergies are deficient in T_{regs}. It seems that without their calming influence, the handbrake is off and the immune system goes full speed ahead, attacking even the most innocuous of substances.

Let me now tell you about cholera – the disease behind the litres of watery white diarrhoea that contaminated the water supply of Soho back in 1854, and which continues to cause outbreaks of misery in developing countries today. It's caused by a nasty little bacterium called *Vibrio cholerae* which colonises the small intestine. But it never intends to stay for long. Where most infectious bacteria attempt to sneak past the immune system until they have fortified their ranks enough to resist attack and cause a persistent infection, *V. cholerae* flaunts its presence from the moment it arrives. In the first phase of its mission, *V. cholerae* attaches itself to the intestinal wall and reproduces as fast as it can. But rather than hang around and cause a permanent infection, this bacterium has other ideas. It gets out. It does so in vast numbers, mixed with the slurry of watery diarrhoea it drains from the body of its host.

Diarrhoea is a strategy of both the bacterium and the immune system. The bacterium uses it as a means to get out and infect further hosts, and the immune system uses it to flush itself clean of a pathogen and its toxins. The mechanism is this: the intestinal wall is rather like a brick wall, in that it is a layer of cells packed closely together. But instead of using a mortar that cements the cells permanently to one another, they are bound by chain-like proteins. This makes the wall a little more flexible. Most of the time anything aiming to get from the gut to the bloodstream is forced to try passing directly through a cell, where it is subjected to all kinds of interrogation. But occasionally, the protein chains can loosen, allowing substances from the blood to the gut, or vice versa. If necessary, water can flood from the blood, between the cells of the gut lining, and into the gut, causing diarrhoea. Useful if the body ever needs to flush out a pathogen.

Two things are remarkable about *V. cholerae*'s watery exit strategy, one of which is pertinent to the impact of our particular microbial colonies and the workings of our immune systems. The other, with which I will start, is just plain interesting.

It concerns the conversation that *V. cholerae* bacteria have with one another before they make their departure. I'm not being anthropomorphic here; these bacteria, and many others beside, really do converse. The infective strategy they have evolved goes something like this: Phase One: infect the small intestine and reproduce like rabbits. Phase Two: when the population has reached a certain size, and the host is on the brink of death, get out in a torrent of diarrhoea, ready to infect ten new hosts. The difficulty with this strategy is how to 'know' when the population is large enough to justify jumping ship. The solution is called quorum sensing. Each bacterium is constantly releasing a tiny quantity of a chemical. In the case of *V. cholerae*, it's called Cholera Auto-Inducer 1, or CAI-1. The more bacteria in the population, the greater the concentration of CAI-1 that surrounds them. Hence, at some stage, a 'quorum' is reached. That is, the threshold minimum number of members are present for voting. With this, the bacteria sense it's time to leave.

The way they make this work is just as slick. The concentration of CAI-1, along with another auto-inducer, AI2, causes *V. cholerae* to change their gene expression. Collectively, they switch *off* the genes that help them to stick to the intestinal wall. They then switch *on* a set of genes that produce substances that force the intestinal wall to open the floodgates. One of these genes codes for a toxin called zonula occludens toxin, Zot for short. It was discovered by the Italian-born scientist and gastroenterologist Alessio Fasano, who is based at the Massachusetts General Hospital for Children in Boston. Zot works by loosening the chains holding the intestinal cells together, allowing water to pour in, and *V. cholerae* to ride their way through the intestines to freedom.

Fasano's discovery got him thinking: if Zot was a viral key that fitted a human lock, what was that lock even doing there? Could there be a *human* key that was meant for that lock, that had been copied by *V. cholerae*? Sure enough, Fasano discovered a new, human, protein just like Zot, which he named 'zonulin'. In the same way as Zot, zonulin interferes with the chains binding the intestinal cells to one another, and so it controls the permeability of the intestinal wall. More zonulin means the chains loosen and the cells move apart, allowing larger molecules to pass in and out of the blood.

Fasano's discovery of zonulin brings us to the bit that's pertinent to immunity. In the presence of a normal, healthy microbiota, the chains are locked in place and the cells are tightly bound. Nothing large or dangerous can get through to the blood. But when the microbiota gets out of balance, it acts a bit like a mild version of *V. cholerae*, irritating the immune system. In response, the immune system tries to defend itself by releasing zonulin to loosen the chains, uncouple the cells of the intestinal walls and flush out the system. The gut lining is no longer an impenetrable wall, keeping out everything but tiny food molecules. Instead, it has grown leaky. Through the gaps between the cells seep all sorts of illegal immigrants, making their way to the promised land of the body.

Now, this takes us into controversial territory. The concept of a leaky gut is a favourite of the alternative health industry, which can be as rapacious and truth-distorting as its more mainstream sibling, Big Pharma. Claims that 'leaky gut syndrome' is the root of all illness, and many other evils beside, are as old as the industry itself. But it has not been until relatively recently that any scientific understanding of its causes, mechanisms or consequences have been scrutinised. Though Big Pharma has many faults of its own, 'alternative medicine' can best be summarised as using two kinds of treatments: those that don't work well enough to merit the title 'medicine', and those that are not yet backed by scientific and clinical evidence. Perhaps a third category of treatment deserves mention: those that cannot be patented and sold, including rest and a good diet.

The scientific and medical establishment are wary of the concept of leaky gut. For patients fed up with not getting answers about persistent fatigue, pain, tummy trouble, headaches and so on, leaky gut offers a believable explanation. It's something that both well-informed, well-meaning alternative health practitioners, and charlatans jumping on the bandwagon, can easily offer their patients as a diagnosis, with a reasonable set of lifestyle recommendations as a cure. There are even tests that patients can take that, when properly administered, measure the degree of 'leakiness', and so alternative health practitioners are able not only to give a measure of the problem, but to track improvements. But for medical doctors, there's little evidence filtering down through health departments and medical schools in support of the concept, and its credibility remains low. The UK's National Health Service website offers little encouragement to patients exploring the idea:

> Exponents of 'leaky gut syndrome' – largely nutritionists and practitioners of complementary and alternative medicine – believe the bowel lining can become irritated and 'leaky' as the result of a much wider range of factors, including an overgrowth of yeast or bacteria in the bowel, a poor diet and the overuse of antibiotics. They believe that undigested food particles, bacterial toxins and germs can pass through the 'leaky' gut wall and into the bloodstream, triggering the immune system and causing persistent inflammation throughout the body. This, they say, is linked to a much wider range of health problems and diseases. The above theory is vague and still largely unproven.

But this view is quickly becoming outdated. An unwillingness to align with the alternative health perspective and risk hard-won credibility is perhaps putting off scientists and doctors who are involved in research regarding intestinal permeability and chronic inflammation.

Even as a science writer, I am hesitant about broaching this topic in a book based rigorously on scientific research, for fear of pushing more sceptical readers out of their comfort zone. But the research base is growing and the mechanisms are being uncovered. Before you make your decision about whether this is quackery or reality, let me review the evidence.

It starts with Alessio Fasano and zonulin. Whilst his intention was to learn more about cholera's invasive tactics, he ended up realising he had answers to another problem, much closer to home. In the 1990s, Fasano was a paediatric gastroenterologist, newly arrived in the States from Italy. Among his patients were children who had coeliac disease. It had been relatively uncommon until then – not even mentioned in a major 800-page report on digestive diseases published in 1994. The children Fasano was treating would become very sick if they ate even a tiny amount of gluten. Gluten is a protein found in wheat, rye and barley grains, that makes bread dough stretch and hold on to the air bubbles produced by yeast. In coeliac disease, this protein brings on an autoimmune reaction. Immune cells treat the protein as if it were an invader, producing antibodies against it. These antibodies also attack the cells of the intestines, causing damage, pain and diarrhoea.

Coeliac disease is quite a special autoimmune disease; it's the only one for which the trigger is known – gluten. This knowledge kept the immunologists happy – they knew what was causing it. Geneticists were also happy: lots of genes had been located that made some people more susceptible to coeliac disease than others. Bad genes plus an environmental trigger equals disease, they thought. But as a gastroenterologist, Fasano wasn't happy. For gluten to cause disease, it had to make contact with immune cells. But to do so it must cross the gut lining. For exactly the same reason as diabetics must inject insulin rather than swallowing it, gluten cannot do this. It is simply too big a protein to cross the gut wall by itself.

But Fasano's discovery of cholera's toxin, Zot, and the human equivalent, zonulin, proved to be the clue he needed. In people who develop coeliac disease, whether they are eight or eighty years old, gluten and a bad set of genes are not enough. Something has to let the gluten through. Fasano knew that the intestinal walls were leaking in coeliacs, and he had a hunch that the zonulin might have something to do with it. He tested the intestinal tissues of children with coeliac disease and children without. As he suspected, the coeliacs had much higher levels of zonulin. Their intestinal walls were opening up and letting gluten proteins through into the blood, where they triggered an autoimmune reaction. Now, about 1 per cent of the population in the West suffer from coeliac disease.

Coeliacs aren't the only ones with leaky guts and high levels of zonulin. Type 1 diabetics also have particularly permeable intestines, and Fasano found that zonulin was behind it once again. In a breed of rats that are used for studies into diabetes, the leaky gut always precedes the disease by several weeks, suggesting it's a necessary step in the development of this autoimmune disease as well. Giving the rats a drug to block zonulin's action prevents two-thirds of them from getting diabetes. So what about the other twenty-first-century illnesses? Do any of them also show leaky guts or increased levels of zonulin?

We'll begin with obesity. I mentioned in chapter 2 that weight gain was associated with high levels of a compound called lipopolysaccharide, or LPS, in the blood. You can think of these molecules as skin cells for bacteria. They kept the bacteria's innards in, and keep the threats of the outside out. They also slough off like skin cells, renewing themselves constantly. LPS molecules coat the surface of 'Gramnegative' bacteria. This is a grouping that has more to do with microbe identification and function than it does with the species, genera and phyla I have referred to so far. Both Gram-negative and Grampositive bacteria live in the gut, and neither is inherently 'good' nor 'bad'. But the fact that the LPS molecules from Gram-negative bacteria are found in the blood in high levels in obese people *is* 'bad'. It begs the question, how did they get there?

LPS is a relatively large molecule. Under normal circumstances, it cannot cross the gut lining. But when intestinal permeability rises – that is, the gut grows 'leaky' – LPS sneaks between the cells of the gut wall and into the blood. On the way, it triggers receptors – guards positioned to ensure there's no breach in the barrier. Their response on encountering LPS is to alert the immune system. They release chemical messengers called cytokines, which whizz around the body raising the alarm and rousing the troops.

In the process, the whole body can become inflamed. Immune cells called phagocytes flood the fat-storing adipose cells, forcing them to get larger and larger instead of dividing into two. Up to 50 per cent of the volume of these cells in an obese person can be not fat, but phagocytes. The bodies of overweight and obese people are in a state of low-level, but chronic, inflammation. Not only does it encourage weight gain, the LPS in their blood also interferes with the hormone insulin, prompting type 2 diabetes and heart disease.

Excess LPS in the blood has also been linked to mental illness. Depressed patients, autistic children and schizophrenics often have leaky guts and chronic inflammation. Intriguingly, traumatic events, from being separated from your mother as a baby to losing a loved one, can turn the gut leaky. Whether this is the missing biological link between suffering stress and developing depression is not yet clear, but as with the gut–microbiota–brain axis, the evidence for it is growing. Depression often accompanies ill-health, from obesity to IBS and acne, but is usually attributed to the misery of the disorders themselves. The idea of a leaky gut leading to chronic inflammation and kick-starting both physical and mental health problems is an exciting one for medical science.

Leaky gut is certainly not the cause of every illness, let alone the political and social ills some would like to blame on it. But the concept may need a rethink, and a rebrand, in the face of the scepticism it currently incurs. Good-quality scientific work into its importance in the genesis of a number of conditions is currently overshadowed by its sullied past. Obesity, allergies, autoimmune diseases and mental health conditions all show significant rises in the permeability of the intestines, with chronic inflammation ensuing. This inflammation comes in the form of an overactive immune system, reacting to the illegal immigrants crossing the gut's border into the body: from food molecules such as gluten and lactose to bacterial products such as LPS. Sometimes the body's own cells get caught in the cross-fire, resulting in autoimmune diseases. A balanced and healthy microbiota seems to act as a gatekeeping force reinforcing the integrity of the gut and protecting the sanctity of the body.

Not only are allergens and the body's own cells in the firing line, but also certain members of the microbiota, as appears to be the case with the most ubiquitous of diseases of civilisation: acne. In the name of scientific research, I have visited some of the most isolated places on the planet. Much of my time away from the cool, wet metropolis of London has been spent in habitats and cultures utterly different from my own. Jungles in which people hunt possums or mouse deer for their dinner. Deserts where the quickest mode of transport is a camel. Communities whose entire villages float on rafts in the sea. In each, daily life is different from life at home. Food is caught, killed and eaten, no supermarkets or packaging required. Nightfall brings darkness, but for an oil lamp or wood fire. And illness is accompanied by the very real possibility of imminent death. These are places where a sleeping child might lose an eye to a chicken, where a workplace accident would be falling out of a tree whilst collecting honey, and where no rain means no dinner. Most fundamentally, meals are what you can grow or catch, and health care relies on a herb and a prayer.

What you don't see in the highlands of Papua New Guinea, tens of kilometres from the nearest road, or the sea gypsy villages of Sulawesi in Indonesia, are people with acne. Not even teenagers. Yet in Australia, in Europe, in America, in Japan, everybody has it. I say everybody, and to the nearest approximation that is true. Over 90 per cent of people in the industrial world, at some point in their lives, suffer from spots. Teenagers get the worst of it, but in the last few decades it seems to have spread beyond them. Now adults, women especially, continue to have acne into their twenties and thirties and sometimes beyond. Around 40 per cent of women between twenty-five and forty years old have some degree of acne, and many of these never even had it as a teenager. More people visit the dermatologist about acne than any other skin condition. Like hay fever, we tend to see acne as just part of life, especially for teenagers. But if that were so, why don't people in pre-industrial parts of the world suffer from it too?

With a little thought, it is plainly ridiculous that so many people have acne. What's even more ridiculous is that so little research has been done into what causes it, despite an inexorable rise in cases, particularly among adults who have long since emerged from pubertal torment. We have stuck with the same old explanation for more than half a century: overactive 'male' hormones, too much sebum, a frenzy of *Propionibacterium acnes*, and hence an ugly immune response of redness, swelling and white blood cells (pus). But on careful inspection it doesn't quite make sense. Women with higher levels of androgens – the male hormones thought to be responsible for acne – don't actually have worse acne. And men, who have far higher levels of androgens, don't suffer from acne as much as women.

So what's going on? New research suggests we have been looking in the wrong place. The idea that *P. acnes* causes acne is decades old, and comes from an obvious source. Want to know what's causing spots? Then look inside a spot and see what microbes linger there. No matter that those same bacteria live on healthy skin in acne sufferers, and on the skin of people without acne. No matter that some spots contain no *P. acnes* at all. The density of *P. acnes* doesn't correlate with the severity of the acne, and levels of sebum and male hormones don't predict the presence of acne.

The fact that antibiotics, both when applied directly to the face and when swallowed as pills, often improve acne has held up the *P. acnes* theory and continues to do so. Antibiotics are the most commonly prescribed drug for spots, and many people remain on them for months or years. But antibiotics don't only affect the bacteria living on the skin. Those in the gut are also affected. We saw earlier that antibiotics change the immune system's behaviour. Could this be the real reason for their action against acne?

What's becoming clear is that *P. acnes* is not crucial to the development of acne. Exactly what role this skin bacterium plays is still up for debate, but new ideas are emerging about the immune system's contribution to this modern skin condition. The skin of acne patients contains extra immune cells, even in apparently healthy areas. It seems that acne is yet another manifestation of chronic inflammation. Some even suggest that the immune system has become hypersensitive to *P. acnes* and perhaps other skin microbes, treating it no longer as a friend, but as a foe.

The same goes for inflammatory bowel diseases (IBD) – Crohn's disease and ulcerative colitis. Possibly because of a shift in the normal composition of the microbiota, the immune cells of the gut seem to lose their normal respect for the gut's colony. It might be because the T_{reg} cells, with their calming influence, have given up on controlling the more aggressive members of the immune platoon. So, instead of tolerating and encouraging beneficial microbes, they mount an attack against them. It's not so much *auto*immunity, directed against the self, but *co*immunity – immune attacks upon the 'commensal' microbes that normally live in a mutually beneficial relationship with our bodies.

The fact that sufferers of IBD are considerably more likely than healthy people to go on to develop colorectal cancers hints at a deeper link between dysbiosis and health. It has long been known that certain infections can promote cancer. The human papillomavirus (HPV), for example, is behind most cases of cervical cancer, and the bacterium *Helicobacter pylori* that causes stomach ulcers can also initiate stomach cancer. The dysbiosis that comes along with IBD seems to add an extra layer of risk. The inflammation it causes somehow damages the DNA of the human cells lining the gut wall, allowing tumours to develop.

The microbiota's role in causing cancer is not limited to cancers of the digestive system. Because dysbiosis can promote a leaky gut and inflammation, cancers of other organs are also promoted by an unhealthy microbiota. Liver cancer provides the clearest example of how this happens. In an experiment to work out how obesity and a high-fat diet might be involved in cancer, researchers exposed both lean and obese mice to carcinogenic chemicals. The lean mice mostly resisted cancer, but one-third of the obese mice developed cancer of the liver. Unsure how a high-fat diet could promote cancer outside the gut, the researchers compared the contents of the blood in the two sets of mice. The obese mice had higher levels of a harmful compound called deoxycholic acid (DCA), which is known to damage DNA.

DCA comes from bile acids – substances produced to help in the digestion of fats in the diet. But it's only in the presence of a particular group of microbes, belonging to the clostridia, that bile acids are converted into DCA, which must then be broken down by the liver. The obese mice had far higher levels of clostridia in their guts than the lean ones, making them especially susceptible to developing liver cancer. Targeting the clostridia in the obese mice with antibiotics reduced the chance they would get cancer.

We all know that smoking and drinking put us at risk of developing cancer, but far less well-known is that we are significantly more likely to get cancer if we are overweight. In men, it's estimated that about 14 per cent of cancer deaths are connected with being overweight, and in women that figure is even higher, at 20 per cent. Many cases of breast, uterine, colon and kidney cancer are thought to be associated with being too fat, and the 'obese' microbiota is at least partly to blame.

The great irony of health in the twenty-first century, now that the reign of infectious disease is over, is that being healthy now may depend on having *more* microbes, not fewer. It's time to move on from the hygiene hypothesis to the Old Friends hypothesis: it is not infections that we're lacking, but the beneficial microbes that train and calm our developing immune systems.

In chapter 1, I asked what links seemingly unrelated twenty-first-century illnesses, from obesity to allergies, and autoimmune diseases to mental health conditions? The answer is the undercurrent to all of them: inflammation. Our immune systems, far from taking a holiday now the reign of infectious diseases has come to an end, are more

FIVE

Germ Warfare

In 2005, Jeremy Nicholson, Professor of Biological Chemistry at Imperial College London, put forward a controversial idea: that antibiotics were behind the obesity epidemic. Fredrik Bäckhed's early experiments showing that the microbiota played a huge role in harvesting and storing energy from food had opened scientists' minds to the possibility that weight gain might be controlled by microbes. If gut microbes could make mice gain weight, could altering the composition of those microbes with antibiotics be playing a part in human obesity?

Though it was not until the 1980s that really large numbers of people became overweight and obese, the trend towards today's obesity epidemic began in the 1950s. Nicholson wondered whether the timing of the introduction of antibiotics for public use in 1944 – just a few years before the inexorable rise in cases of obesity began – was more than mere coincidence. His suspicions were not simply down to chronological correlation. He already knew that farmers had been using antibiotics for decades to fatten their animals for market.

In the late 1940s, scientists in the US had made the accidental discovery that giving chickens antibiotics boosted their growth by as much as 50 per cent. Times were hard, and an increasingly city-dwelling American public were fed up with the high cost of living. They were beginning to feel they had gone without for long enough, and cheaper meat was high on their post-war wish-list. The effects of antibiotics in chickens seemed nothing short of miraculous, and farmers rubbed their hands with glee when they discovered that cattle, pigs, sheep and turkeys all had the same response to small daily doses of the drugs – a massive increase in growth.

Though they had no idea how these drugs promoted growth, or what the consequences might be, food was short and prices were high. Production gains of this magnitude for the cost of, well, chicken feed, were spectacular. Ever since, so-called subtherapeutic antibiotic therapy has been an essential part of farming. Estimates are vague, but it's possible that up to 70 per cent of America's antibiotics are used in livestock. The added bonus that more animals can be crammed into a smaller space without succumbing to infections only furthered their use. Without these growth promoters, the US would need to breed an extra 452 million chickens, 23 million cattle and 12 million pigs each year to produce the same weight of meat.

Nicholson's concern was this: if antibiotics could make livestock considerably fatter, what proof did we have that they weren't doing the same to us? The human digestive system isn't all that different from a pig's. Both pigs and humans have an omnivorous diet, a simple stomach, and a large colon filled with microbes that make use of the leftovers of our own digestive process in the small intestine. Antibiotics can boost piglet growth rates by roughly 10 per cent per day. For farmers, this means they are ready for slaughter two or three days sooner – a big gain across thousands of animals. Is it possible that humans are also being fattened for market through our voracious consumption of antibiotics?

For many people who struggle with their weight, being slim is their greatest desire, yet no matter how much they want it, they cannot achieve it. So powerful is this longing that morbidly obese patients who lost significant amounts of weight said, in one study, they would rather lose a leg or be blind or deaf than return to being obese. Every single one of the forty-seven patients said they would prefer to be slim than be an obese multi-millionaire.

If people so desperately want to be slim, why is it so easy to put on weight and so difficult to lose it and keep it off? Even the most optimistic estimates suggest only 20 per cent of overweight people manage to lose weight and keep it off for more than a year at some point in their lives. People who have managed to lose weight report having to eat far fewer calories to stay slim than a typical weight-maintenance diet for their height suggests they should. Losing weight is so difficult that some government-led advice has now abandoned trying to get people to drop the pounds; they simply want us to avoid getting any fatter. America's 'Maintain, Don't Gain' campaign follows this rubric; many workplaces now offer courses and counselling to help their staff avoid weight gain during traditional times of indulgence, such as holidays and over Christmas.

It's a suspicion that tallies with our evolving understanding of obesity as an illness. If, as I explained in chapter 2, Professor Nikhil Dhurandhar is correct in thinking that obesity is not a simple matter of an imbalance of calories-in versus calories-out, but rather a complex disease with a great many possible causes, then antibiotics could prove to be an important factor in the epidemic. It would provide an attractive explanation for some of the extraordinary data associated with obesity. Even the mere fact that 65 per cent of people in some developed countries are overweight or obese presents a mind-boggling view of human behaviour. Are we really so lazy, so greedy, so ignorant and so unmotivated that more members of our species have become overweight than those who remain lean? Or are there deeper reasons for our excess weight than we have assumed?

An obesity epidemic induced or encouraged by antibiotics would not only absolve us of at least some responsibility for our excessive weights, but could also give us a means of beating it without resorting to seemingly futile diets.

In 1999, New York-born ex-nurse Anne Miller died at the age of ninety, having lived fifty-seven years longer than she should have. In 1942, when she was thirty-three, Miller suffered a miscarriage. She then developed a streptococcal infection that had her hours from death in a hospital bed in Connecticut. As Miller's temperature approached 42°C (107°F), her doctor asked her family for permission to take drastic action in a bid to save her life.

The doctor wanted to try a new drug, never used in a patient before, that he had heard was being developed by a pharmaceutical company in New Jersey. Its name was penicillin. Miller had been delirious with fever for a full month when, on 14 March at 3.30 pm, she was injected with a single teaspoonful of the drug – half of the world's total supply. By 7.30 pm her fever had subsided and her condition had stabilised. A few days later, she had fully recovered. Anne Miller's life was the first ever to be saved by antibiotics.

Since then, antibiotics have prevented the deaths of countless millions of people, beginning in earnest with Second World War soldiers wounded as they crossed Normandy's beaches on D-Day in 1944. As stories of miraculous recoveries reached the public, demand for the drug grew. By March 1945, the pace of penicillin production had picked up, and in the US, anybody who wanted it could purchase a course through their local pharmacy. By 1949, the price had dropped from US$20 for 100,000 units of penicillin, to just 10 cents. In the sixty-five years since then, a further twenty varieties of antibiotics have been developed, each targeting bacteria in a different way. Between 1954 and 2005, production of antibiotics in the United States shot up from 900 tonnes per year to 23,000 tonnes. Amongst them, these amazing drugs have changed the way we live and die. Their invention is one of humanity's greatest triumphs, preventing suffering and death at the hands of our oldest and deadliest enemy. It's hard to imagine now, but these were once miracle drugs. Their use was reserved for the most desperate cases, and they were, quite literally, life-savers.

Now, for better or worse, their use is pandemic. I challenge you to find a single adult living in the developed world who has not taken antibiotics at least once in their life. In Britain, the average woman will take seventy courses of antibiotics during her lifetime. Seventy. That is not too far off one course every year. The average man, perhaps due to his innate reluctance to visit the doctor, or maybe because of differences in male and female immune systems, will take fifty courses. In Europe, 40 per cent of people have taken antibiotics within the past twelve months. In Italy, that figure is 57 per cent, balanced out by lower proportions in Sweden (22 per cent). Americans match the Italians – around 2.5 per cent of them are on antibiotics at any given moment.

Indeed, even finding a child under the age of two who has not re-ceived a course of these drugs will prove difficult. Around one-third of children are given antibiotics by the age of six months, rising to nearly half at age one, and three-quarters at age two. By the time they turn eighteen, children in developed countries will have received between ten and twenty courses of antibiotics on average. Around a third of all antibiotics prescribed by doctors are consumed by children. America's youth get through 900 courses of antibiotics for every 1,000 children each year. Spanish children are taking a whopping 1,600 courses per 1,000 children per year – that is, on average, they are given 1.6 courses of antibiotics every year of their young lives.

About half of these childhood prescriptions are for ear infections, which young children are particularly prone to. The little tube con-necting the ear to the throat – the one that 'pops' with pressure changes – is nearly horizontal in babies, and steepens with age. This means that in infants, mucus does not drain into the throat as easily, and the tube tends to clog with debris. Ear infections are also around twice as com-mon in children given a dummy to suck, which is by now practically ubiquitous. These infections are taken seriously by doctors because of two risks – both of which are very small: one, young children with repeated infections can sometimes have difficulty hearing, at a time crucial for them to learn to speak; and two, these infections can turn nasty if they spread deeper and affect the mastoid bone behind the ear. Known as mastoiditis, this bacterial infection can cause perma-nent hearing damage or even death. Although both of these risks are extremely small, it's enough to persuade many doctors to stay on the safe side.

Predictably enough, not all of this drug-taking is strictly necessary. America's public health institute – the Centers for Disease Control and Prevention – estimates that half of the antibiotics prescribed in the US are unnecessary or inappropriate. Many of these prescriptions are for people suffering from colds and flu, desperate for a cure, and granted by doctors too weary to deny them a placatory offering. No matter that both colds and flu are the work of viruses, not bacteria, and antibiotics can't touch them. Or that the majority of colds will burn themselves out in days or weeks, without risk to life or limb.

As antibiotic resistance becomes an ever more serious problem, the pressure is on doctors to be judicious in their prescribing habits. There's plenty of room for improvement. In the US in 1998, three-quarters of all the antibiotics doled out by primary care doctors were for five respiratory infections: ear infections, sinusitis, pharyngitis (sore throat), bronchitis and upper respiratory tract infections (URI). Of the 25 million people who went to their doctor about a URI, 30 per cent were prescribed antibiotics. Not so bad, you might think, until you realise that only 5 per cent of URIs are caused by bacteria. The same goes for sore throats; 14 million people were diagnosed with pharyngitis that year, and 62 per cent of them were given antibiotics. Only 10 per cent of them would have had bacterial infections. Overall, around 55 per cent of antibiotic prescriptions given out that year were unnecessary.

As gatekeepers to pharmaceutical drugs, it might seem as if the responsibility for the overuse of antibiotics ultimately lies with doctors, but patient ignorance can be an overwhelming pressure. In a Europe-wide survey of 27,000 people in 2009, 53 per cent wrongly believed that antibiotics killed viruses, and 47 per cent believed they were effective against colds and flu – which are caused by viruses. For many doctors, the fear of sending a sick patient away empty-handed, only for them to return with a serious complication of a bacterial infection, is enough to persuade them to prescribe antibiotics just in case. Babies, especially, can fill an inexperienced doctor with dread: an infant might be screaming for want of a cuddle, or because of serious pain. It might be quiet and listless because of a large dose of calming Calpol, or because it is critically ill. For a young doctor, safe is better than sorry. But is it worth it?

In some cases, it is. Chest infections, for example, turn out to be pneumonia relatively often, especially in older people. For one older person with symptoms to avoid pneumonia through antibiotics, about forty people will be treated without any benefit. But for many other illnesses, antibiotics are wasted on far more patients for every one that avoids a serious complication. Over 4,000 people with sore throats and URIs will receive antibiotics without cause, in order to avoid complications in just one person. The risk is even smaller for children with ear infections. It has been estimated that around 50,000 children would need to be treated with antibiotics to prevent a single case of mastoiditis. Even so, most children with mastoiditis recover uneventfully, and the risk of death is around one in ten million. The antibiotic resistance that will develop as a result of treating all those children is undoubtedly more dangerous to public health than the tiny risk of the infection itself.

It's clear that in the developed world, we take vast quantities of antibiotics, and most of them are unnecessary. The contrast with the use of antibiotics in the developing world, where infectious diseases remain common and antibiotics save lives, was brought to life in a BBC Radio 4 interview with Professor Chris Butler, who is a practising family doctor as well as Professor of Primary Care at Cardiff University in Wales. He says:

> I originally came to the UK from a large rural hospital in South Africa where the burden of infectious diseases was incredible, and otherwise fit, healthy people were coming in with pneumonia and meningitis in large numbers. They were on death's door and if we gave them the right antibiotic in time, in a few days they often got up and walked out of there. We were making the dead get up and walk through this miracle drug of antibiotics.

When I came to the UK I started working in general practice. Here, we were using the same antibiotic in UK general practice that had saved so many lives in South Africa, to treat what were effectively snotty-nosed kids.

So why not take antibiotics just in case? What harm can they do? Butler's concerns about using life-saving drugs to placate mildly ill patients were based mainly around the development of antibiotic resistance. He, like many other scientists and doctors, predicts we may soon enter a post-antibiotic era much like the pre-antibiotic era, in which surgery carried a high risk of death, and minor cuts could kill. This prediction is as old as antibiotics themselves. Sir Alexander Fleming, after making his discovery of penicillin, repeatedly cautioned that using too little of it, for too short a time, or without good reason, would bring about antibiotic resistance.

He was right. Time and time again, bacteria evolve resistance to antibiotics. The first penicillin-resistant bacteria were discovered just a few years after penicillin was introduced. It's as simple as this: the susceptible bacteria die, sometimes leaving behind those that have, by chance, a mutation that makes them resistant. Then, the resistant bacteria reproduce and the entire population is immune to the antibiotic. In the 1950s, the common bacterium *Staphylococcus aureus* had become resistant to penicillin. Some members of the species had a gene that produced an enzyme called penicillinase, which breaks down penicillin, making it ineffective. As all the bacteria that did not have the penicillinase gene were killed, those that did became dominant.

A new antibiotic – methicillin – was introduced in the UK for treating penicillin-resistant *Staph. aureus* infections in 1959. But just three months later, a new strain of *Staph. aureus* emerged in a hospital in Kettering. Resistant not only to penicillin but to methicillin too, it is now known as the dreaded MRSA – methicillin-resistant *Staphylococcus aureus*. MRSA kills tens or hundreds of thousands of people each year, and it is not the only antibiotic-resistant bacterium.

The consequences are not just societal, but personal as well: 'We know that the biggest risk factor for getting a resistant infection is that if you've recently had antibiotics.' As Chris Butler says:

> So if you have had antibiotics, when you get your next infection the chances of that being resistant are much greater. And that is a problem because even for common infections like urinary tract infections, if they're caused by a resistant organism they go on for longer, people get more additional antibiotics, it costs more to the NHS and people have symptoms that are much worse. So it's not just harming the future sensitivity of bacteria, there's also a big downside for individuals who take unnecessary antibiotics.

Antibiotic resistance, though, appears not to be the only downside of our overuse of antibiotics. Chris Butler brings up a further worry: harmful side effects. He and his team ran a large clinical trial, testing the benefits of antibiotics for people with a sudden-onset cough.

> We found that we had to treat thirty people to get one person to benefit in terms of avoiding a new or worsening symptom, but at the same time, for every twenty-one people treated, one was harmed. So you see that the numbers needed to treat to achieve benefit is more or less evenly balanced against the numbers needed to harm through antibiotic treatment.

That harm comes most often in the form of skin rashes and diarrhoea.

In the seventy years since the introduction of penicillin, an additional twenty classes of antibiotics have been developed, each targeting bacteria in a different way. They are among the most commonly prescribed drugs, and efforts continue to find even more antibiotic compounds to fight the ever-changing threat from evolving bacteria. However, our triumph over our greatest natural adversary – bacteria – has taken place in ignorance of the collateral damage that antibiotics have been causing along the way. For these potent drugs do not only destroy the bacteria that make us sick, but also those that keep us well.

Antibiotics cannot target a single strain of bacterium. Most are 'broad-spectrum' – they kill a wide range of species. That's pretty useful for doctors, as it means people can be treated for all sorts of infections without knowing exactly which kind of bacterium is causing the problem. Being more precise would mean culturing and identifying the culprit, which is slow, expensive, and sometimes impossible. Even the more targeted, 'narrow-spectrum' antibiotics do not select only the disease-causing bacterial strain for destruction. Any other bacteria belonging to the same family group will suffer the same fate. The consequences of this mass bactericide run deeper than anyone ever predicted, Sir Alexander Fleming included.

Both of these antibiotic downsides – resistance and collateral damage – combine forces in one dreadful condition: *Clostridium difficile* infection. This bacterium, better known as *C. diff*, became a serious concern in England in 1999, when it killed 500 people, many of whom had been treated with antibiotics. In 2007, nearly 4,000 people died for the same reason.

It's not a nice way to die. *C. diff* lives in the gut, where it produces a toxin that causes relentless, foul-smelling, watery diarrhoea. With it comes dehydration, horrendous abdominal pain and rapid weight loss. Even if *C. diff*'s victims avoid kidney failure, they may also have to survive toxic megacolon. A megacolon is just as it sounds – the excess gases produced in the gut cause the colon to swell up far beyond its normal size. As with appendicitis, the risk is that it will burst, but the consequences are even more deadly. With faecal matter and bacteria of all kinds released into the sterile environment of the abdominal cavity, the odds of survival drop significantly.

C. diff's rising incidence, and death toll, stems partly from antibiotic resistance. In the 1990s, *C. diff* evolved a dangerous new strain that grew more and more common in hospitals. It was both more resistant, and more toxic. But there was an underlying cause as well, that puts our overuse of antibiotics into a stark and frightening light.

C. diff hangs around in some people's guts, not causing too much trouble, but doing no good either. Give it the slightest opportunity, though, and *C. diff* will turn nasty. It is antibiotics that bring about these opportunities. Normally, when the gut microbiota is healthy and balanced, it keeps *C. diff* in check, crowding it out and restricting it to small pockets where it cannot do any damage. With antibiotics though, especially broad-spectrum ones, the normal microbiota is disturbed and *C. diff* can gain a foothold.

If antibiotics can allow *C. diff* to flourish, then the question is: does taking antibiotics change the composition of the microbiota? And if it does, how long does the effect last? Most people will have experienced the bloating and diarrhoea that often seem to accompany a course of antibiotics. It is the most common side effect of antibiotic treatment, and of course it is the result of disruption to the microbiota – dysbiosis. Usually, it clears up within a few days of finishing treatment. But what of the microbes left behind? Do they regain a normal healthy balance?

A team of researchers in Sweden asked this very question in 2007. They were particularly interested in what happens to *Bacteroides* species, as these bacteria specialise in digesting carbohydrates from plants, and, as we've seen in chapter 2, they have a big impact on human metabolism. The researchers split a group of healthy volunteers into two, and gave half of them the antibiotic clindamycin for seven days, and the other half no treatment at all. The gut microbes of those given clindamycin were dramatically affected immediately after receiving the drugs. The *Bacteroides*, in particular, sharply declined in their diversity. The microbiota of both groups were assessed every few months, but by the end of the study the *Bacteroides* of the clindamycin group had still not returned to their original composition. Their treatment had finished two years previously.

Just five days of the broad-spectrum antibiotic ciprofloxacin, which can be used to treat urinary tract infections and sinusitis, has a similar effect. It has a 'profound and rapid' impact on the microbiota, altering the composition of species in the gut within just three days.

The bacterial diversity drops, and around a third of the groups present change in numbers with treatment. This shift persists for weeks, and some species completely fail to recover. The consequences of antibiotics on babies can be even more dramatic. In a study looking at changes to the microbiota of infants as they age, a course of antibiotics for one baby left it with so few bacteria that the researchers couldn't even detect any DNA.

This long-term impact on the gut microbiota has been seen for at least half a dozen of our most common antibiotic drugs, each changing the composition in different ways. Even the shortest of courses and the lowest of doses can have an impact that far outlives the original complaint. Perhaps this doesn't seem so bad – after all, change doesn't have to be for the worse. But think back to the rise of twenty-first-century illnesses. Type 1 diabetes and multiple sclerosis in the 1950s, allergies and autism in the late 1940s. The obesity epidemic has been blamed on the introduction of self-serve supermarket shopping and the associated guilt-free pleasures of anonymous consumption. But when did they proliferate? The 1940s and 50s. These dates coincide with another important event: Anne Miller's last-minute retreat from the grim reaper's dark embrace. Or, more accurately, the D-Day landings in 1944, when antibiotics became available in large quantities.

Their use on this historic day was quickly followed by the release of antibiotics for public use. Syphilis was the prime target, affecting, as it did, around 15 per cent of adults at some point in their lives. Before long, antibiotics were cheap to produce and frequently prescribed. Penicillin remained a firm favourite, but five other classes of antibiotics were introduced within a decade, each targeting a different bacterial weakness.

For some twenty-first-century illnesses, a slight delay between 1944 and the spark point of their rise might suggest a mismatch. But this lag is to be expected. It took time for antibiotics to reach common usage, for further antibiotic drugs to be developed, for children to grow up with the influence of these drugs on their bodies, and for chronic diseases to develop in their own insidious way. It also takes time for the effects to become clear across populations, countries and continents. If the introduction of antibiotics in 1944 is in some way responsible for our current state of health, the 1950s are exactly when we would expect to see the dawning of their impact.

Let us not jump the gun though. As any scientist would hasten to point out, correlation does not always mean causation. The timely introduction of antibiotics may be as unrealistic a connection to rising chronic illness as the self-serve supermarkets that made their debut in the 1940s. Connections alone, whilst useful guides, do not always provide a *causal* link. An amusing website about spurious correlations tells me that there's an impressively close correlation between per capita consumption of cheese in the US and the number of people who die each year by becoming tangled in their bed sheets. Cheese-induced nightmares aside, it's pretty unlikely that eating cheese causes death by tangling, or that death by tangling leads others to consume more cheese.

Finding a true causal link requires two things. First, evidence that the connection is real. Does taking antibiotics really entail a greater risk of developing a twenty-first-century illness? Second, a mechanism for how the first causes the second. *How* does taking antibiotics bring on allergies, autoimmunity or obesity? The idea that antibiotics might change the microbiota, which changes metabolism (obesity), brain development (autism) and the immune system (allergies and autoimmune diseases), needs backup from more than just the synchronous timing.

Quite aside from the impact of antibiotics on the weights of farm animals, we have known since the 1950s that antibiotics can bring about weight gain *in humans*. At that stage – before the obesity epidemic began – antibiotics were actually used intentionally in humans for their growth-promoting properties. Some pioneering doctors, aware of the newly discovered impact of antibiotics on the growth of livestock, tried treating premature and malnourished babies with antibiotics. For the newborns, the results were spectacular – they benefited from a boost in weight that apparently lifted them from the clutches of mortal danger. But from the perspective of today's overweight masses, these trials should perhaps have sounded a warning.

The results at that point were not confined to youngsters. A trial of antibiotic treatment among US Navy recruits in 1953 started out as a test of whether prophylactic aureomycin could reduce the burden of streptococcal infections. That the heights and weights of these young men were recorded was down to typical military thoroughness, but it led to a surprising observation. The recruits receiving the antibiotics were gaining significantly more weight than those receiving an identically packaged placebo. As with the babies, this unexpected outcome of antibiotic treatment was viewed in terms of the potential nutritional value it might offer, rather than an alarming indicator of times to come.

With the obesity epidemic in full swing, and our new appreciation of the microbial dinner party we host in our guts, these early experiments provide an alternative view of the role of antibiotics in weight gain. It is shocking that such a clear effect in animals, and even in humans, has been neglected as a research avenue, given the proportions of the obesity epidemic. We knew antibiotics caused weight gain, because we made use of them for beefing up our livestock and enhancing the nutrition of those who need it most. Yet we ignored this side effect in the context of a global health catastrophe.

The discoveries of Bäckhed, Turnbaugh and others that I talked about in chapter 2, combined with Nicholson's prediction, have prompted a fresh look at this old connection. Clearly the microbiota plays a big role in weight gain, but do antibiotics switch it from a lean community to an obese one?

It's a difficult question to answer: giving large numbers of otherwise healthy people antibiotics to see if they get fat is hardly ethical, so scientists must rely on natural experiments, and the odd mouse or two. Researchers in Marseille in France recognised the opportunity to test Jeremy Nicholson's theory using adults who were suffering from a dangerous infection of the heart valves. These patients desperately needed copious quantities of antibiotics to get better, providing a great opportunity to look for associated weight gain. The researchers compared changes in their Body Mass Index (BMI) over the next year with those in a healthy set of people who were not given antibiotics. The patients gained far more weight than the healthy people, but only those on one particular combination of antibiotics – vancomycin + gentamycin – were affected. Those on other varieties of antibiotics were spared, gaining no more weight than the healthy people.

By looking at the gut microbiota of these two groups, the researchers could see whether any particular species might be responsible for the weight gain. They discovered that one species, called *Lactobacillus reuteri* (a member of the Firmicutes phylum), was far more abundant in the guts of the patients who had been given vancomycin. This bacterium is resistant to vancomycin, meaning it could spread like a weed whilst other species were felled by the drugs. Through germ warfare, antibiotics had given an unfair advantage to the enemy. Not only that, but *L. reuteri* produces antibacterial chemicals of its own. Known as bacteriocins, these compounds can prevent regrowth of other bacteria, ensuring that *L. reuteri* retains its dominance of the gut. *Lactobacillus* species like *L. reuteri* have been given to livestock for decades: they too make animals gain weight.

Another study tapped into the informational treasure trove of the Danish National Birth Cohort. Researchers analysed health data for nearly 30,000 mother–child pairs and found that giving babies antibiotics had different effects, depending on the weight of their mothers. Giving antibiotics to the children of slim mothers made them *more* likely to become overweight. But giving them to the children of overweight and obese mothers had the reverse effect: a reduced risk of becoming overweight. It's hard to know for sure why antibiotics are causing the opposite effect in these babies, but it's intriguing to think that perhaps they are 'correcting' an obese microbiota on the one hand, and disrupting a lean one on the other. In another study, it was found that 40 per cent of overweight children had been given antibiotics in their first six months, whereas only 13 per cent of normal weight children had.

Persuasive though these experiments are, they don't prove that antibiotics can cause weight gain, or that it's a consequence of an altered microbiota and not a direct effect of the drugs themselves. A team led by New York University's Martin Blaser – an infectious disease doctor and director of the Human Microbiome Project – set out to determine precisely what antibiotics could do to the microbiota and metabolism. In 2012, they had shown that giving low doses of antibiotics to young mice disrupted their microbiota composition, changed their metabolic hormones and increased their fat mass, though it did not alter overall weight gain. They suspected that timing was critical, and that giving the mice antibiotics from a younger age might have a more substantial effect. Epidemiological studies had shown that human babies given antibiotics in the first six months were more likely to become overweight than those who made it past their first birthday without being exposed to the drugs. The same went for farm animals – to get the best weight-boosting effect, it's important to start providing antibiotics as early as possible.

In a second set of experiments, Blaser's group tried giving low doses of penicillin to pregnant mice just before they gave birth, and then continuing the drugs throughout nursing. Predictably enough, male mice on penicillin grew much faster than controls whilst they were being nursed. As adults, both males and females were heavier, with a greater fat mass than mice which had not received any drugs.

The team were keen to know what would happen if the mice not only received low-dose penicillin, but also ate a high-fat diet. In females, mice on a normal diet developed around 3 g of fat within their little bodies by the age of thirty weeks, regardless of whether they were given penicillin or not. Feeding an identical set of female mice a high-fat diet pushed their fat mass up to about 5 g – they weren't heavier, but they had a lower lean mass and a higher fat mass. But adding a low dose of penicillin to that high-fat diet caused a further group of female mice to develop a fat mass of not 5 g, but 10 g. The penicillin had somehow amplified the effect of eating an unhealthy diet, making the mice store even more of the calories they were taking in.

For male mice, eating unhealthily had a greater effect, taking their fat mass from 5 g on a normal diet (with or without penicillin) to 13 g on a high-fat diet. Once again, combining the high-fat diet with low-dose penicillin amplified the gain in fat, boosting the fat mass of the male mice to around 17 g. Clearly, a high-fat diet alone was causing obesity in the mice, but antibiotics were making a bad situation far worse.

Transplanting the altered community of microbes that had been produced by the low-dose antibiotics into germ-free mice induced the same changes in weight and fat, indicating that it was the microbial composition that caused weight gain in the mice, and not the drugs themselves. Worryingly, although stopping the antibiotics allowed the microbiota to recover, the metabolic effects of the treatment lingered on. Penicillin is the most commonly prescribed antibiotic class in children, and if mice are anything to go by, treatment with these drugs early enough in life could cause permanent changes to their metabolisms.

It's too early to be certain that antibiotics can cause obesity, or to know which varieties might be responsible, but given the vast, and growing, scale of the obesity epidemic, these tantalising suggestions of a cause deeper than greed and sloth should put us on our guard against overusing these precious and complicated drugs. Martin Blaser warns that between 30 and 50 per cent of American women are routinely given antibiotics during pregnancy or labour, mostly penicillins such as his mice received. Though many of these are undoubtedly warranted, it's clearly important that we reassess the costs and benefits of this practice as new evidence emerges.

As for the farm animals, evidence of the transfer of antibiotic resistance from livestock to humans has put a stop to the use of antibiotic growth promoters, in Europe at least. Since 2006, farmers in the European Union have been banned from using antibiotics purely to boost weight, though they are still allowed for curing disease, of course. In the US and many other countries, antibiotic growth promoters continue to be used on a daily basis. It's enough to make you wonder, even if you manage to avoid putting antibiotics in your own body, are you inadvertently drugged when you tuck into a steak, or through the milk on your cereal? After all, many antibiotics are absorbed into the blood and on into the muscle or milk of an animal, and some may even survive cooking and reach your gut when you eat them. Fortunately for those living in the most developed of countries, there are strict rules forcing farmers not to slaughter or milk recently treated animals. On the other hand, in many less strict countries, spot checks often reveal food that is contaminated by antibiotic residues beyond safe levels. Depending on where you live, and where you've travelled, chances are you have absorbed at least some antibiotics through your food.

The vegans amongst us might be counting themselves lucky after reading that, but despite their virtues they are not excluded from this particular threat. Vegetables, though not dosed with antibiotics directly, are often grown in soil fortified with animal manure. Manure is not only a rich source of nutrients, but also of drugs – around 75 per cent of antibiotics given to animals pass straight through them. So much for using nature's fertiliser to keep things clean and green. For some types of antibiotics, there can be as much as one dose of the antibiotic in every litre of manure. That works out as the equivalent of sprinkling the contents of one or two capsules of antibiotics over every 10 m^2 of farmland.

Some of these antibiotics remain 'active' once in the soil – as able to kill bacteria as if they were in their packets. This means the concentration can rise even further with every application of manure. All of this would not matter if the antibiotics remained in the soil, but they don't. Vegetables and herbs from celery to coriander to corn contain antibiotic residues – tiny amounts per stick or sprig or tinful, but over weeks and years the impact adds up. There are rules about giving animals antibiotics before they are slaughtered for the dinner table, but no rules about drugs in manure used for cropland. On your plate of meat and two veg, it may well be that the vegetables pass on the burden of agricultural antibiotics while the meat is spared.

Whether antibiotic residues in food could really be underlying the obesity epidemic is a matter for scientific debate, but the connection is certainly compelling. Our girths began to grow in the 1950s shortly after antibiotics were made available to the masses. But a sharp upturn in the number of overweight people occurred in the 1980s, around the same time there was a shift to super-intensive farming operations. There are about 19 billion chickens alive on Earth at any given moment (that's nearly three per person!), many of which are squeezed into multi-layered cages. To keep chickens in these conditions without them getting sick usually involves a lot of antibiotics. Public health specialist Dr Lee Riley points out that in the Eighties and Nineties the biggest increase in these antibiotic-reliant chicken farms occurred in the south-eastern states of America – exactly where the epicentre of the obesity epidemic lies, and where Americans are now the fattest.

If antibiotics may be capable of making us fat, what other damage might they be responsible for? I've mentioned several other conditions that appear to be associated with dysbiosis in the gut: allergies, autoimmunity, and several mental health disorders. As antibiotics can disrupt the microbiota, each of these illnesses could, in theory, be brought on by treatment with these drugs.

Remember Ellen Bolte from chapter 3, whose son Andrew developed autism as a toddler? Bolte blamed Andrew's sudden regression on the repeated courses of antibiotics the little boy was prescribed for what seemed to be an ear infection. Like obesity, autism is another condition which was once rare. It has been on the rise since the Fifties, and now affects 1 in every 68 children. Boys are worst hit, with nearly 2 per cent of them considered to be on the autistic spectrum by the age of eight. The blame has been placed on many things, the most controversial of which was the combined measles, mumps and rubella (MMR) vaccine. Evidence for a causal link for this was absent though, and research attention has turned to the microbiota.

Children with autism seem to harbour an unbalanced community of microbes. This dysbiosis affects a toddler's developing brain, making them irritable, withdrawn and repetitive. Could Ellen Bolte have been right about the cause of Andrew's autism? Antibiotics are clearly capable of disrupting the microbiota, but were Andrew's antibiotics really responsible for making him so unwell? His diagnosis with constant ear infections provides a clue. It turns out that 93 per cent of children with autism had ear infections before they turned two, compared with 57 per cent of children without the condition. As I mentioned, no doctor wants to leave a childhood ear infection alone, lest it stops a toddler from learning to speak, or leads to something nasty such as rheumatic fever. So they turn to antibiotics – better safe than sorry.

The link between more ear infections and more antibiotics bears up. An epidemiological study showed that kids with autism tend to have been given three times more antibiotics than those without it. Those getting antibiotics under the age of eighteen months appear to be at the greatest risk. What's more, this is a real connection. It is not something we can blame on hypochondriac parents, convinced their child is ill and hence requesting antibiotics or pushing for an autism diagnosis. Nor can we blame it on generally poor health that has led to more illnesses as well as autism. We know that neither of these is the case, because before their autism diagnosis, the autistic children in this study hadn't visited the doctor or taken more of any other medications than normally developing children. Studies of greater numbers of children are needed to be sure of the link, and a clear mechanism explaining how this might happen is also crucial before we can be certain. But given that we are already being cautioned to reduce antibiotic use, the threat of ramping up the risk of autism only makes these warnings more pertinent.

A much clearer, and perhaps more intuitive, link can be seen between antibiotics and allergies. I mentioned in the previous chapter that children treated with antibiotics before the age of two are twice as likely to go on to develop asthma, eczema and hay fever than those who weren't. The more drugs they received the more likely they were to become allergic; four courses or more meant three times the likelihood of developing allergies.

The plot thickens when we consider autoimmune diseases. They too have risen in line with the use of antibiotics, but until recently, the blame for these conditions has been laid on infections. Type 1 diabetes is a classic. For decades, doctors have seen a pattern: a teenager comes to see them with a cold or flu, then weeks later is back, desperately thirsty and unbearably tired. The beta cells of their pancreas have begun to pack in, refusing to release insulin any more. Without this crucial hormone to convert and store it, glucose in the blood builds up. It sucks water into the kidneys, dehydrating the unfortunate teenager. Within days or weeks, things can get serious, leading to coma or death without treatment. But what's intriguing is the connection that has been made between getting a cold or flu and getting diabetes. A viral infection is often seen as the trigger, and not just for diabetes, but for many other autoimmune conditions.

Looking at the statistics, though, reveals something different. The risk of getting type 1 diabetes is not higher in children who have actually had infections. What's more, while cases of type 1 diabetes have been rising by about 5 per cent a year in the United States, rates of infectious illness have declined. So why the apparent link? Why do doctors consistently see teenagers develop diabetes after they have had an infection?

Here's where the science ends and the intrigue begins. We already know that doctors overprescribe antibiotics, even for illnesses that are probably viral, not bacterial. Could it be that diabetes begins not because of an infection, but as a result of the treatment for that infection – antibiotics? To families and to doctors, it would seem as if a cold, or the flu, or a bout of gastroenteritis kicked off the diabetes. Antibiotics would appear to be an innocent bystander, but it could be that it's the drugs themselves, or a combination of the two, that triggered the diabetes.

Unfortunately, for now the answer is not clear. In a Danish study of antibiotics given to young children, there was absolutely no link to the risk of developing diabetes later. But in another study of over 3,000 children, there was a tendency for antibiotic use and diabetes to be linked. Diabetes aside, other autoimmune conditions show clearer connections with antibiotics. Among teenagers and adults using an antibiotic called minocycline for months or years to keep acne at bay, the risk of developing lupus was two and a half times greater than for those not taking this drug. This autoimmune disease, which attacks many parts of the body, mainly affects women. But this figure included men, who aren't at all prone to getting lupus. Looking at just women, the risk of getting lupus after taking minocycline (but not other tetracycline antibiotics) shoots up to around five times greater risk than for those who take no antibiotics. So too for multiple sclerosis (MS) – an autoimmune condition that damages the nerves – which is more likely to hit those who have recently taken antibiotics. Whether antibiotics, infections or a combination of the two are at the root of this is hard to know.

Although the problems of antibiotic resistance and collateral damage to the microbiota are serious ones, antibiotics are not all bad. Let's not forget the countless lives they have saved and the suffering they have prevented. In recognising that they have costs as well as benefits, we can better assess their value in any given situation. It's the job of all of us – including both doctors and patients – to reduce our use of *unnecessary* antibiotics, for the good of our inner ecosystems and our own bodies.

Although the idea behind the hygiene hypothesis – that infections protect us from allergies – has turned out to be false, there's an element of it which lives on. We are, as a society, obsessed with *hygiene*, and through its impact on the beneficial microbes that we harbour, we are coming to harm. Most of us in developed countries wash our entire bodies at least once a day, covering our skin with soap and hot water. It's often said that the skin is the first line of defence against pathogens, but this is not quite true. The skin's microbiota, whether it's a community of *Propionibacterium* on the nose, or one of *Coryne-*

bacterium in the underarms, forms an added layer of protection on the skin's surface. As in the gut, this beneficial layer crowds out potential pathogens and regulates the immune system's responses to would-be invaders.

If antibiotics can dramatically alter the composition of the gut microbiota, what effect are soaps having on the skin microbiota? Looking on the shelves of supermarkets now, it's hard to find a hand-wash or surface cleaner that doesn't contain antibacterials. We are besieged by advertising implying that killer germs are on the loose in our homes and telling us to keep our families safe by using cleaning products containing antibacterials that kill 99.9 per cent of bacteria and viruses. What they don't tell us in these adverts is that normal soap does just as good a job, and it won't harm you or the environment in the process.

When you wash your hands properly, with warm water and non-antibacterial soap, you are not getting rid of potentially harmful microbes by killing them. You are physically removing them. The soap and the warmth of the water do not harm them, they simply make it easier to remove the substances the microbes are clinging to – meat juices, fats, dirt, or your skin's own build-up of oils and dead cells. The same goes for surface cleaners – cleaning the kitchen counter removes the leftover bits of food that harmful bacteria can feed on. It doesn't actually kill the microbes, and it doesn't need to. Adding antibacterials achieves nothing extra.

When antibacterial products claim to kill 99.9 per cent of bacteria, they are not referring to tests on people's hands, or on kitchen surfaces, but in pots. Testers put a large number of bacteria directly into the liquid soap, and after a period of time – far longer than the soap would be in contact with your skin – they see how many are still alive. Claiming a 100 per cent kill rate is impossible, because no one can ever prove a complete absence of anything from a small sample. As scientists say: absence of evidence is not evidence of absence. Exactly which *strains* of bacteria these soaps kill is rarely declared; the 99.9 per cent refers to the proportion of individuals killed, not that 99.9 per cent of the world's bacterial species can be eliminated. It's worth bearing in mind that many pathogenic bacteria can form spores anyway, effectively hibernating until the danger has passed, regardless of the chemicals used.

Antibacterial products are a triumph of advertising and assumptions over science. Like many chemicals in our everyday lives, the safety of the antibacterial chemicals has never really been scrutinised. Rather than requiring chemicals to be proven safe and effective *before* they are sold, like pharmaceutical drugs, it's up to the regulatory agencies to prove *after* they have been unleashed on the public that they are dangerous, in order to get them banned. Of the 50,000 or more chemicals in use in the West, just 300 or so have actually been tested for safety. Five have been restricted in their use. That's 1.7 per cent of those tested. If we assume just 1 per cent of the other 50,000 are harmful, that's 500 more that shouldn't be in our homes.

It's easy to be blasé about their impact – after all, wouldn't we see people getting sick if these chemicals really were dangerous? – but the insidious nature of chemical accumulation, and the subtle, slow-burn effects they might have, are not necessarily that easy to pick up. Besides, we *are* seeing people get sick. Our memories are short enough, and the web of potential factors tangled enough, that we have real difficulty teasing apart what's dangerous and what's not. Take asbestos, for example. This naturally occurring chemical was used in construction all over the world before being banned. Hundreds of thousands of people have died from exposure to this once-ubiquitous chemical, and they continue to do so.

I'm not suggesting that antibacterials are as acutely dangerous as asbestos, but that the fact that they are in thousands of products, from cleaners to chopping boards, towels to clothing, plastic containers to body washes, doesn't guarantee that they are safe. One particularly common antibacterial compound, known as triclosan, has come under scrutiny in recent years. Its effects are sufficiently worrying that the governor of Minnesota has signed a bill banning its use in consumer products from 2017. I'm almost certain you'll have at least one product containing triclosan in your home, if not dozens. But you'd probably be better off without it.

For starters, triclosan has been shown to be no more effective at reducing bacterial contamination in the home than non-antibacterial soap. But while people continue to use it, triclosan is itself contaminating our water supply, where it *does* manage to kill bacteria, and disrupt the balance of freshwater ecosystems. As if this wasn't concerning enough, triclosan also permeates our bodies. It can be found in human fat tissue, in the umbilical cord blood of newborn babies, in human breast-milk, and in significant quantities in the urine of 75 per cent of people on any given day.

How concerning this is is still being thrashed out in the scientific literature, but what we know so far is that there's a clear correlation between the levels of triclosan in people's urine and the severity of their allergies. The more triclosan our bodies contain, the more likely we are to have hay fever and other allergies. Whether this is a direct effect of damage to the microbiota, a form of toxicity, or even a reflection of reduced exposure to beneficial microbes is not known. Whatever it is, it puts a different spin on the hygiene-focused adverts showing mums cleaning their toddlers' high chairs with antibacterial wipes before placing food directly on the 'clean' surface.

There's even evidence to show that triclosan actually makes you *more* likely to get an infection. We are literally dripping with triclosan: it is even to be found in the 'nasal secretions' – snot – of adults. But rather than this nasal impregnation with antibacterials helping us to fight off infections, it's been found that the greater the concentration of triclosan in the snot, the *higher* the colonisation by the opportunistic pathogen *Staphylococcus aureus*. By using triclosan, we are actually reducing our body's ability to resist colonisation, and making it easier for this bacterium, which kills tens of thousands of people each year (in the form of MRSA), to cling on.

As if all this wasn't enough, triclosan has also been found to interfere with the action of thyroid hormones, and to block the action of oestrogen and testosterone in human cells in Petri dishes. For now, the

Food and Drug Administration (FDA) of the United States has simply challenged manufacturers to prove triclosan's safety, or face a ban. As I mentioned, the governor of Minnesota has jumped ahead, banning triclosan in consumer products with effect from 2017, but not for any of the reasons I have just mentioned. His concern, and that of many microbiologists, is that exposing bacteria to triclosan enables them to develop resistance. While nobody wants to wash all the vulnerable beneficial bacteria off their hands, leaving just the harmful, resistant ones behind, the worry is focused on antibiotic resistance. Take one human nose, pumping out possible resistance-inducing triclosan-soaked snot, add some *Staphylococcus aureus* and leave for a few days, and what do you get? A mobile MRSA-factory, complete with a highly effective dispersal mechanism.

Oh, and one more thing. When triclosan is combined with the chlorinated water from the tap, it is converted to the carcinogen and favoured incapacitating agent of crime novelists, chloroform. Wait for the ban if you like, or alternatively: always read the label.

Hand-washing – with normal non-antibacterial soap and warm water for fifteen seconds – *is* important, though. It is the mainstay of public hygiene, and has been shown to make all the difference when it comes to transmitting infections, especially gastrointestinal ones. But, as well as removing 'transient' microbes – the non-resident bugs you pick up from the environment – hand-washing also disrupts the microbiota of your hands. Intriguingly, different species vary in their ability to either resist a wash, or recover soon after it. Members of the staphylococci and the streptococci, for example, form a greater proportion of the community immediately after hand-washing, and then gradually become less dominant between washes.

I say that this is intriguing because it brings to mind obsessive–compulsive disorder (OCD). In one manifestation of this anxiety disorder, sufferers feel that they are contaminated with germs. They develop 'obsessions' with cleanliness, and 'compulsions' to wash their hands. The causes of this bizarre and highly life-limiting affliction are hard to pin down, despite many diverse theories. One set of clues points to a microbial origin.

As the First World War came to an end, a mysterious illness began in Europe. By the winter of 1918 it had reached America, and the following year it hit Canada. Over the coming years, the illness swept the globe, taking in India, Russia, Australia and South America. The pandemic persisted for an entire decade. Known as encephalitis lethargica, its symptoms included extreme lethargy, headaches and involuntary movements, a bit like Parkinson's disease. Often, the disease manifested itself as a psychiatric disorder, with many sufferers becoming psychotic, depressed or hyper-sexualised. Between 20 per cent and 40 per cent of sufferers died.

Many survivors of the encephalitis lethargica pandemic didn't fully recover; thousands were left with OCD. Suddenly, a rare behavioural disorder was cropping up as if it were an infection. The doctors of the day argued fiercely about whether the disease had a Freudian or an 'organic' origin, but it would be another seventy years before its cause was discovered.

In the 2000s, two British neurologists took an interest in the cause of encephalitis lethargica. Dr Andrew Church and Dr Russell Dale had seen a handful of patients whose symptoms fitted the profile of this bizarre disease. Word got around the medical community and colleagues began to refer similar cases to Dale, until he had twenty patients who had all been diagnosed with an illness that had been assumed to have disappeared decades previously. Together with Church, he began to look for similarities between the patients, hoping that they would find clues that could lead them to a cause, and with luck, a treatment. Fortunately, there was a pattern: many of the patients had suffered a sore throat in the acute phase of their illness.

As Americans – who often use the term 'strep throat' – will know, sore throats are very often caused by members of the genus *Streptococcus*. Church and Dale thought they might be on to something with this bacterium. They checked their patients, and sure enough, all twenty of them had been infected with *Streptococcus*. Rather than clearing up after a few weeks, the strep had triggered an autoimmune reaction which attacked a cluster of brain cells known as the basal ganglia. As a result, what would normally have been a respiratory infection became a neuropsychiatric condition.

The basal ganglia are involved in 'action selection' – this part of the brain helps us to decide which of several possible simple actions we should choose to actually carry out. The basal ganglia seem to be able to subconsciously learn what actions will bring us a reward – should you stick or should you twist? Should you brake or should you accelerate? Should you reach for your cup of tea or should you scratch that sudden itch on your head? The more practice you have at each of these things, the more information your basal ganglia have when selecting between the choices running through your conscious mind. Sticking or twisting depends on what cards you have, what cards the dealer has, and what cards your mind has managed to assess remain in the pack. The more practice you have, the more acutely your basal ganglia are tuned, even if your conscious mind is not.

If these brain cells get attacked, though, action selection goes awry. Should you stick or twist? Or twist or stick? Or twist? Or stick? Or just *twitch* with indecision. Muscles that should automatically follow the brain's instructions seem to get multiple orders, and instead of smooth decisive actions, they produce Parkinson-like tremors. Routines are also messed with – turning the lights on, locking the doors, washing the hands. For those OCD sufferers who compulsively wash their hands, there's an intriguing possibility. I mentioned earlier that some groups of bacteria become *more* abundant just after hand-washing, perhaps because they take the opportunity to bloom in the absence of their more vulnerable peers. I'll save you the trouble of looking back: the streptococci are one such group. It's by no means certain, but maybe these opportunistic pathogens gain enough ground on the hands and in the gut after a good hand-wash to persuade their host, via the habit-enforcing, reward-giving basal ganglia, to keep on washing.

Perhaps it won't come as a surprise that a number of 'mental health' disorders – more appropriately referred to as neuropsychiatric disorders – are connected to both basal ganglia dysfunction and to *Streptococcus*. Think of the vocal and physical tics of Tourette's syndrome, which may be the result of the basal ganglia's failure to decide not to suppress the conscious mind's idea of mischief. *Streptococcus* infection plays a role here, making children fourteen times more likely to develop Tourette's if they have suffered multiple infections of a particularly nasty strain of strep in the past year. Parkinson's, ADHD and anxiety disorders are also linked with strep and damage to the basal ganglia.

I'm not suggesting we shouldn't wash our hands, lest strep takes hold, though. A far worse outcome would be to transmit common microbes from places they belong (in stool, for example) to places they don't (the mouth or eyes). Whether antibacterial soap worsens the temporary takeover of *Streptococcus* on the hands is not known, but as a hardy opportunist used to resisting attack by other microbes, it's entirely possible that it's been quicker to evolve resistance to antibacterials than the beneficial microbes of the skin. There is, however, one occasion when using chemicals to kill bacteria seems both worthwhile and effective, and that is alcohol hand-rubs. Alcohol disrupts microbes on such a fundamental level that they appear not to be able to evolve resistance. What's more, alcohol can be effective against antibiotic-resistant strains like MRSA, and can be applied quickly and easily by health-care workers as well as by commuters.

While you're busy checking all the labels on your personal hygiene products, it might strike you just how many chemicals you've never heard of that it seems to take to keep you feeling clean and smelling fresh. Of course, your skin would look after itself without shower gel, moisturiser and deodorant. If traipsing round tropical rainforests teaches you anything, it's that it's the outsiders who wash once a day and cover themselves in anti-perspirants that smell bad, not the local people. Despite washing infrequently, and never using deodorants or cleansers, tribal people living in the most basic of places don't suffer from bad body odour.

Gita Kasthala, an anthropologist and zoologist who works in

remote areas of West Papua and East Africa, has noticed that people in tribal societies can be split into three groups when it comes to personal hygiene. The first are those who have had very little contact with Western culture. 'These people often incorporate personal hygiene into other activities, say when they go fishing. But they don't use soap, and many of the fabrics they use to cover themselves are natural,' she says. The second group are from remote villages that have been exposed to Western culture to some extent – often through religious missionaries – and they tend to wear Western clothing, usually second-hand synthetic fabrics from the Eighties. 'This group often have an incredibly pungent odour. They wash themselves in an ad hoc manner, and will use soap, but have little understanding of why you wash yourself and your clothes. They just know that at some point it should be done, be this weekly, monthly, or more seldom.' The final group have been fully immersed in Western culture, perhaps through working at an oil base or for a logging company, and wash daily with cosmetics. 'This group do not generally smell, unless they heavily exert themselves or it is a hot day,' Kasthala explains. 'But the first group, who never use soap, don't ever smell, even when they exert themselves.'

So why not? How is it that most people living in modern society become socially-unacceptably smelly and greasy after just a day or two of not washing, but those living soap- and hot-water-free in the tropics manage to stay clean?

According to a newly launched company called AOBiome, it's all about a very sensitive group of microbes. The company's founder, David Whitlock, was a chemical engineer who studied the microbes found in soil. While collecting soil samples from a stable yard in 2001, he was asked why horses loved to roll in the dirt. He didn't know, but the question got him thinking. Whitlock knew that soil and natural water sources contained a lot of ammonia-oxidising bacteria, or AOBs. He also knew that sweat contained ammonia, and he wondered if horses and other animals used the AOBs in the soil to manage the ammonia build-up on their skin.

Most of the smell of human sweat doesn't actually come from the ammonia-containing fluids that our eccrine glands release, but from the apocrine glands, or scent glands. These glands, which are confined to the underarms and the groin, are all about sex. They are not even active until puberty, and the smells they produce after puberty act as pheromones, informing the opposite sex about our health and fertility status. But the sweat that the apocrine glands release is actually completely odourless. It only gains a scent when our skin microbes tuck in and convert it into a whole host of smelly volatile compounds. Exactly which scents are produced depends on the composition of the microbes you harbour.

By washing and applying deodorants, which tend to work by removing or masking the odour-producing bacteria, we alter the skin microbiota. AOBs are a particularly sensitive group of bacteria, and they are slow to repopulate, so they are the worst-affected by the hurricane of chemicals that bombards them every day. The trouble is, according to Whitlock, without AOBs, the ammonia we sweat out is not converted into nitrite and nitric oxide, chemicals which play some fundamental roles not only in regulating the running of human cells, but also in governing our skin microbes. Without nitric oxide, the corynebacteria and staphylococci that feed on our sweat can run wild. Changes in the abundance of corynebacteria in particular seem to be responsible for the bad body odour that we are all so keen to avoid.

The irony is, then, that in washing with chemicals and using deodorant to ensure that we smell nice, we kick off a vicious cycle. Soap and deodorant kill off our AOBs; no AOBs means disruption to our other skin bacteria; the altered composition means our sweat smells unpleasant; and therefore we must use soap to clean up the mess, and deodorant to mask the smell. What AOBiome suggest is that replenishing our AOBs could break this endless loop.

Of course you could achieve this by rolling in the mud, or swimming in some unpolluted, untreated water each day (if you can find it), but what Whitlock and the rest of the team at AOBiome propose is that you spray yourself daily with their AO+ Refreshing Cosmetic Mist instead. It looks, smells and tastes just like water, but it contains live *Nitrosomonas eutropha* – AOBs cultivated from soil. At the moment, AO+ is sold as a cosmetic, so AOBiome aren't required to prove its effectiveness – that's their next aim. But in a pilot trial, volunteers had improved skin appearance, smoothness and tightness compared to those using a placebo.

While people who don't wash wouldn't smell as floral or soapy as we have come to expect from our skin, many of the volunteers on the AO+ trial found that their natural smell was actually just as nice, even to other people. AOBiome founder David Whitlock gave up washing entirely twelve years ago, and we are assured that he doesn't smell. Many other members of the AOBiome team have also cut down on their use of body soaps and deodorants, and most wash only a few times a week or even a few times a year.

The idea of not washing with soap, or at least doing so less often, probably seems pretty disgusting to most people. It actually strikes me as surreal that it is so ingrained in our culture that it's practically taboo to admit to not using soap every day. It's probably far more surreal that after not washing with soap for 250,000 years of our history as *Homo sapiens*, we are now so reliant on a daily soapy shower that we can't imagine life without it.

Like antibiotics, antibacterials have their place. But, in health, your body is not that place. We already have a microbial defence system: it's called the immune system. Perhaps we'd do well to try using it.

SIX

You Are What They Eat

As I sit down with Dr Rachel Carmody for a cup of tea at Harvard University, she tells me about the moment she realised we were looking at the human diet in completely the wrong way. She had just finished her Master's dissertation on the effect of cooking on the nutritional value of food, and was presenting her work in her viva examination. At the end of the meeting, the examiner sitting at the far end of the long table from Carmody stood up and slid a stack of newly published scientific papers towards her. As they fanned out in front of her, Carmody caught sight of 'microbiome' and 'intestinal microbiota' among the titles. 'You might want to have a think about how this would affect your conclusions,' the examiner said.

'The amount of energy that we can extract from the diet drives all of biology,' Carmody explains. 'Chances are that the way an organism looks and behaves has to do with how it procures its food. The trouble was, as a human evolutionary biologist looking at how humans digest their food, I was studying only half of the problem.' Carmody had been focused on the digestive processes going on in the small intestine, but as prestigious journals like *Nature* and *Science* began to publish special issues about the role of the gut microbiota in nutrition and metabolism, she realised that her research, and that of others studying human nutrition, could never yield all of the answers. 'What we had, Carmody tells me, 'was a system of thinking about the diet that was distressingly incomplete.'

Our entire perspective on nutrition has shifted. Until recently, what went on in the small intestine was all that mattered. This long, thin tube that leads off the blender-like stomach is where it all happens, as far as 'human' digestion is concerned. Enzymes pumped in from the stomach, pancreas and the small intestine itself break down the large food molecules into smaller ones that are capable of crossing through the cells of gut lining into the bloodstream. Proteins, like twisted and folded pearl necklaces, are cut into single beads, called amino acids, and shorter chains of these building blocks. Complex carbohydrates are sliced into more manageable chunks called simple sugars, such as glucose and fructose. And fats are pruned into their parts: glycerols and fatty acids. These smaller units go about their business in the body, making energy, building flesh and being repurposed for our own use.

Human nutrition, according to dogma at least, essentially stops at the end of the seven-metre-long tube that is the small intestine. It's followed by the much shorter, but wider, large intestine, but this relatively mucky section of bowel has been neglected up until now – as if it's simply an outsize waste pipe. At school, many of us were taught that, while the small intestine was there to absorb nutrients, the large intestine was supposed to absorb water, and to gather up the leftovers of our food ready to be egested. Like the not-so-pointless appendix, the large intestine's importance has been overlooked. The Russian Nobel prize-winner Elie Metchnikoff, who made great discoveries of immune cells in the last decade of the nineteenth century, thought we would all be better off without a large intestine. He wrote: 'very many investigations that have been made in the case of man seem to have established the absence of digestive power in the large intestine.'

Fortunately, we have come a long way since Metchnikoff's musings on the large intestine. We recognised decades ago that, at the very least, this organ also absorbed crucial vitamins, synthesised by its colony of microbes. Without them, our health would suffer. The

'Bubble Boy', David Vetter, who was kept nearly germ-free in isolation in the 1970s, had his diet supplemented with several vitamins to make up for his lack of microbes. But the nutritional contribution of the microbiota goes way beyond vitamins. Indeed, for some species, eating would be almost completely futile if it weren't for the nutrient-extraction services provided by their microbes.

Blood-sucking leeches and blood-lapping vampire bats take their nutritional reliance on their microbiota to the extreme. For all its life-giving properties, blood is not the most nourishing of foodstuffs. It is high in iron, of course, hence the metallic taste, and also in protein, but provides little in the way of carbs, fats, vitamins or other minerals. Without their gut microbes to synthesise these missing elements, such sanguivorous species as leeches and vampire bats would struggle to survive.

The giant panda is another classic. Though it is a Carnivore with a capital C – meaning it belongs alongside its fellow bears – the grizzly, the polar and others – in the tree of life, as well as with the lions, wolves and other such ferocious cousins, it is not a carnivore in its lower case sense – a meat-eater. Having shunned meat for the dietary delight that is bamboo, the giant panda has turned its back on its evolutionary past. Its simple digestive system lacks the great length of more accomplished herbivores such as cows and sheep. This leaves the large intestine and its microbial inhabitants with a lot of work to do.

What's more, pandas have the genome of a carnivore – lots of genes that code for enzymes capable of breaking down the proteins that make up meat, but none that code for enzymes to break down tough plant polysaccharides (carbohydrates). Each day, the panda munches its way through around 12 kg of dry, fibrous bamboo stalks, but only 2 kg of that gets digested. Without its microbiota, that 2 kg would dwindle to nearly nothing. The giant panda microbiome, however – that is, the genes contained within the panda's microbiota – carries a suite of cellulose-busting genes that are more normally found in the microbiomes of herbivores, including cows, wallabies and termites. With these genes – and the microbes that carry them – on its side, the giant panda has escaped from the constraints of its meat-eating past.

So, you see, the microbiota are not to be ignored when it comes to nutrition. In some senses, humans are no different from leeches, bats or pandas. Some of the foods we eat are digested by the enzymes coded for by our own genome, and then absorbed in the small intestine. But many food molecules – mostly the 'indigestible' ones – are left over. These go on to the large intestine, where they meet a large and eager crowd of microbes, ready to break them down using their own enzymes. In the process of feeding themselves, the microbiota release another set of leftovers. These molecules, along with the water we were told about at school, are absorbed into the blood. They are, as we will discover, more important than we ever anticipated.

As for Carmody, she completed her PhD knowing that she had answered only half of the question she had asked, then walked the short distance from Harvard's Human Evolutionary Biology building to the Center for Systems Biology across the way, to embark on answering the other half of the question under the guidance of the seasoned microbe hunter Peter Turnbaugh.

I'm afraid the obesity-related get-out-of-jail-free card I gave you in the last chapter needs to be handed back. The consequences of taking medicinal antibiotics, and ingesting them with our food, are certainly worth attention, especially in young children. But we are not off the hook just yet. As Martin Blaser's work on low-dose penicillin and a high-fat diet in mice suggests, antibiotics are unlikely to be the *sole* cause of weight gain, or of the other twenty-first-century illnesses. Diet, too, has a role to play. It's just not the role you might expect.

The way we eat has changed. Wander the aisles of a supermarket with half a mind on where food comes from – plants and animals – and you will be struck by how little of it actually looks like plants or animals. For starters, half the aisles are filled with food inside cardboard packets, and plastic bags and bottles. How many plants or animals come in cardboard boxes? Broccoli? Chickens? Apples? Granted, sometimes raw ingredients do get packaged this way, but I am thinking of the biscuits, the crisps, the soft drinks, the ready meals and the breakfast cereals. Some American supermarkets look to the outsider less like food shops and more like warehouses. Being unfamiliar with the brands makes it as hard to decipher what foods the rows of boxes contain as it is for an English-speaker to order sushi from a menu written in Japanese. Real food is recognisable. An apple is an apple. A chicken is a chicken. I suspect a shopper from the 1920s would have no need for half the aisles of a modern supermarket.

Outside the supermarket the change continues. Fast food and ready meals for those too busy to cook. Cans of fizzy drinks and bottles of juice for when water isn't tasty enough, takeaway meals for Friday nights, ready-made sandwiches for the office. Most of us are no longer in full control of what we eat. We cook our meals ourselves less often than ever before, and we very rarely *grow* it ourselves. Almost all food, both from plants and animals, is intensively produced, chemically enhanced and spatially restricted. Even those of us who think our diets are healthy are probably comparing them with the average – a pretty low bar.

But what *is* a healthy diet? Advice on this seems to change as often as most people change their shoes, but one thing is quite clear: something about the modern diet is not working out well for us. The number of overweight and obese people worldwide has reached such proportions that it would be better termed a pandemic than an epidemic. Aside from obesity, poor diet also makes a whole host of diseases worse, from arthritis to diabetes. Poor diet is responsible for the majority of deaths in the developed world, be it from heart disease, stroke, diabetes or cancer. Many of the conditions brought on or worsened by a poor diet are twenty-first-century illnesses, including irritable bowel syndrome, coeliac disease and a whole host of conditions related to being overweight.

The trouble is, no one seems able to agree on what constitutes the best diet. Proponents of different popular diets are almost evangelistic about the benefits of their chosen strategy, even though several of the most common approaches take vastly different stances on which food groups are 'good' and which are 'bad'. It seems there is an evolutionarily logical explanation behind every popular diet, as well as claims of weight lost and health gained to back them all up. Perhaps simply taking control of your diet is enough to produce an improvement, regardless of whether you try low carbs, low fat or low GI.

The proposed theologies of most popular diets today come down to one basic tenet: that this is how humans are *meant* to eat. Having utterly lost sight of what constitutes a normal human diet, we turn to our ancestors. Pre-agriculturists, hunter-gatherers, cavemen and even raw-food-munching prehumans, more reminiscent of our cousins among the great apes than our modern selves. But perhaps we don't need to look so far back. After all, our great-grandparents didn't suffer the maladies we complain of now, and they grew up only 100 years or so ago. Certainly, in comparison with the cardboard-filled supermarkets and drive-through eateries of the present day, their approach to eating was drastically different from our own.

One way to get a good idea of 'what humans are meant to eat' is to look to populations whose lives have not been permeated by intensified farming, globalised feeding and convenience-based dietary choices. The population of the rural village of Boulpon in Burkina Faso, for example. It was the children of this African village who a group of Italian scientists and doctors chose to compare with a group of children living in urban Florence in Italy. Their aim was to learn about the effect that diet had on the gut microbiota of each of these two groups. The people of Boulpon lead lives not so different from those of subsistence farmers living around 10,000 years ago, just after the Neolithic Revolution.

This was a decisive period in human evolution. Humanity had hit on two important ideas: domesticating animals, and intentionally growing and tending food crops. Aside from the obvious benefits of a stable food supply, the Neolithic Revolution enabled groups of humans to give up their nomadic lifestyle and settle in one place. With that came the opportunity to build permanent structures, live in larger groups, and unfortunately to sustain infectious diseases within their populations. It marked the beginnings of our modern culture and our modern diet.

For the diets of our ancestors, agriculture and animal husbandry meant a steady supply of grains, beans and vegetables, and in some locations, eggs, milk and an occasional bit of meat. In some places, little has changed. In the village of Boulpon, the local diet is typical of rural Africa: cereal grains of millet and sorghum ground into flour and mixed into a savoury porridge, dipped into a sauce of locally grown vegetables. Now and then a chicken is killed and eaten, and during the rainy season, termites sometimes make for a tasty treat. Not exactly a meal plan fit for a best-selling Western what-we're-meant-to-eat diet book, but probably more representative of our recent ancestral diet than a typical meat-feast hunter-gatherer-style fad diet. The Italian children, meanwhile, were eating foods common in a modern, Western diet: pizzas, pasta, a lot of meats and cheeses, ice cream, soft drinks, breakfast cereals, crisps and so on.

Not surprisingly, these two groups of children had totally different gut microbes. Where the Italian kids mainly had bacteria belonging to the Firmicutes group, the Burkina Fasan children's microbiotas were dominated by bacteria in the Bacteroidetes group. Over half of the microbes in the guts of Burkina Fasan children belonged to a single genus, *Prevotella*, and a further 20 per cent were from the genus *Xylanibacter*. But these two genera, apparently so important to the African children, were completely absent from the guts of the Italian children.

What strikes you about the difference in the diets of the Italian and Burkina Fasan children? Probably, it's the amount of fat and sugar (simple carbohydrates) that the Italian children are eating in comparison with the children from Burkina Faso. We all think that we eat too much of these food types, and that, above all else, they are responsible for the obesity epidemic and associated diseases. Certainly, the quickest way to make a lab rat fat is to feed it a high-fat, high-sugar diet. It's what microbiome scientists call 'the Western diet'. After just a single day eating this way, rats' gut microbes have changed composition and are making use of different sets of genes. Within two weeks of switching to the Western diet, mice and rats become fat.

So is the opposite also true? Can returning to a low-fat or a low-carbohydrate diet reverse the microbial changes and bring about weight loss? Ruth Ley, the woman who first noticed the high Firmicutes to Bacteroidetes ratio in obese humans, wanted to know whether the ratio would shift back to that found in lean people if overweight people went on a diet. So she used weights and faecal samples from obese human volunteers enrolled in a diet trial to find out.

The volunteers were assigned either a low-carbohydrate or a low-fat diet to follow for six months. Their weights were recorded and their gut microbiotas sampled before and during the experiment. Both groups of dieters lost weight over the first six months, and their Firmicutes to Bacteroidetes ratios fell in line with the proportion of their own body weight that they had lost. Intriguingly, this microbial shift was only evident once the dieters had lost a certain proportion of their weight. For those on the low-fat diet, a 6 per cent weight loss was needed before the relative abundance of Bacteroidetes began to reflect their efforts. That's a 12 lb weight loss for an obese woman of 5′6″ with a starting weight of 200 lb. For those on the low-carb diet, just a 2 per cent weight loss was enough to begin to impact the microbial ratio – a 4 lb weight loss for the same obese woman.

Whether this discrepancy in the initial weight loss that was needed to push the microbiota towards a leaner balance is meaningful in just twelve volunteers remains to be seen, as does the importance of the microbial ratio itself. But it's interesting to note, because low-carb diets are famed for their quick results, although over longer periods of dieting, low-fat diets catch up and sometimes overtake low-carb diets in terms of total weight lost.

In Ley's experiment, though, both diets were also low in calories, allowing women 1,200–1,500 calories per day and men 1,500–1,800 per day. The fact is, eating a diet that's low enough in calories for a significant period of time will always result in weight loss, regardless of whether they are low in fat, low in carbs, only allow you to indulge at the weekend, or leave out the grains and dairy products of the Neolithic Revolution. Even balanced-nutrient diets, which reduce neither relative fat content nor relative carb content, result in weight loss, as long as caloric intake is kept low. Ley and others now believe the Firmicutes to Bacteroidetes ratio may correspond to dietary habits more than it reflects obesity itself. If any low-calorie diet will do when it comes to reducing weight, does it actually make sense to restrict either fat or carbohydrate intake?

Making sweeping statements that fat and carbohydrates are 'bad' belies the great complexity of these foods. Just like claiming that cars are bad because they kill people, but ignoring the fact that they ease our lives, suggesting that fat is always bad overlooks the fact that it is crucial to survival. I can't claim to know better than anyone else what the best balance of saturated, mono-unsaturated, polyunsaturated and trans fats consists of. As they are with fad diets, opinions on which of these will harm your health and which are safe are polarised, even among the experts.

As for the hitherto overlooked microbiota's response to fat and sugar, it's very hard to assess even in experimental conditions. Imagine you want to use mice to test the impact of a high-fat diet on gut microbes. You supplement their normal chow with extra fat and see what happens. But now your mice are receiving far more calories than before, so you don't know if the changes are down to the increase in fat or the increase in calories. So instead, you increase their fat intake but keep their caloric intake steady by reducing the amount of carbohydrate in their chow. The trouble is, now you don't know whether any changes are because of the increase in fat or the decrease in carbohydrate. Nothing in nutrition can be studied in isolation.

As I said, putting mice on a high-fat but low-carbohydrate diet causes a shift in the microbial composition, and a gain in weight. Alongside these changes come increases in the permeability of the gut wall, the amount of lipopolysaccharide (LPS) in the blood, and markers of inflammation. These changes have been associated not just with obesity, but with type 2 diabetes, autoimmunity and mental health conditions. A diet high in simple sugars such as fructose appears to create the same broad changes, at least in rodents.

So it seems too much fat and sugar are bad for you, and the rise in their consumption parallels the rise in obesity in many countries around the world. But here's the paradox: in the UK, parts of Scandinavia and Australia, in contrast to the popular media image of a morbidly obese person clutching a burger and a super-sweet milkshake, consumption of fat and sugar has actually *fallen* since the Second World War. In the UK, the government-led National Food Survey tracked food intake in British households from 1940 to 2000. The statistics go against every assumption we make about changes in our diets over time. For example, in 1945, the average fat content of the British diet was 92 g per person per day. In 1960, when very few Brits were overweight, it was 115 g per day. But by 2000, it had dropped to 74 g per day. Even breaking down fat into its different fatty acids – saturates and unsaturates – fails to offer an explanation. Fatty acids traditionally thought of as better for us – the unsaturates – form an ever-greater part of the British fat intake. Butter, whole milk and lard are down, and skimmed milks, oils and fish are up. Yet Brits continue to gain weight.

Looking at the relationship between fat intake (as a proportion of total energy intake) and BMI reveals no relationship either. Across eighteen European countries, average BMI and average fat intake were entirely unrelated in men: the proportion of fat that men ate and their weights were not connected. In women, the relationship was the opposite of what you might expect: countries with higher average fat intakes (up to 46 per cent of the diet) had lower average BMIs, and those with lower fat intakes (as low as 27 per cent of the diet) had higher BMIs. Eating more fat doesn't necessarily make you fatter.

Sugar consumption is harder to judge because the survey collected data about food types, but not sugar content itself. But consumption of table sugar, jams, and cakes and pastries has also fallen over time. In the case of table sugar, consumption went from a whopping average of 500 g per person per week in the late Fifties to around 100 g per person per week in 2000. However, Brits now drink far more fruit juice and eat more sugar-loaded breakfast cereals than they used to. Overall, estimates suggest that since the 1980s, Brits consume about 5 per cent less sugar – roughly a teaspoon's worth per day. Since the 1940s, bearing in mind that this includes a period of rationing during and after the Second World War, the decline has probably been much greater. Australians, too, have reduced their sugar intake since the Eighties. Whereas they used to get through around 30 teaspoons a day in 1980, in 2003 they had dropped to 25 teaspoons each per day. Over this same time period, three times as many Australians became obese.

An increase in overall calorie consumption doesn't seem to account for the changes either. The UK's National Food Survey shows that in the 1950s, average daily energy intake for adults and children combined hit 2,660 calories, but by 2000 it had dropped to 1,750 calories per person per day. Even in America, some studies show that caloric intake has declined during periods of overall weight gain. The United States Department of Agriculture (USDA) figures from their Nationwide Food Consumption Survey show a drop in energy intake from 1,854 calories to 1,785 calories per person per day between 1977 and 1987. At the same time, fat intake fell from 41 to 37 per cent of the diet. Meanwhile, the proportion of people that were overweight rose from a quarter to a third of the population. A simple excess in calories-in versus calories-out does seem to contribute at some points in time, in some places, but many scientists have pointed out that it can't explain the sheer scale of the obesity epidemic, in America or elsewhere.

My point is not that fat and sugar are not bad when consumed to excess – they can be. And not that consumption of one or both hasn't risen overall across the world as a whole – it probably has. Rather it is that, as in experimental mouse diets, an increase in the consumption of one nutrient *must* impact on the consumption of other nutrients, especially if overall caloric intake remains level. The great debate in recent years has been whether it is fat or sugar causing the obesity epidemic. But what if it is neither?

If changes in fat, sugar and calorie consumption don't fully explain the rise in obesity, what does? As ever more people have become overweight, we have asked, what has risen in our diets to explain this? The answer seemed obvious: fat and sugar. In many places, an increase in consumption of both paralleled the increase in obesity.

But whilst overconsumption of fat has an intuitive appeal as the precedent to weight gain, it is not as simple as it seems. We see the excess fat on our own bodies, and liken it to the excess fat we see in our food: the fatty rind on the edge of a piece of steak, or the white streaks running through rashers of pink bacon. But this is illogical. Fat on our bodies can be created from any food source that needs to be stored – protein, carbohydrate or fat.

Although we instinctively feel that there should be, there isn't actually a big difference in fat intake in the diets of the children in Burkina Faso and those in Italy. Children in Boulpon consumed about 14 per cent of the diet as fat, and Florentine children about 17 per cent. What if looking for an increase in an element of our diets was too simplistic? What if we should also have looked at what has *decreased*?

It's not quite as instinctive to associate a fall in one thing with a rise in another, but it's just as plausible. Looking again at the diets of the children from Burkina Faso in comparison with those from Italy, we can see one clear difference in nutrient intake: fibre. The vegetables, grains and beans that make up the bulk of the Boulpon diet are all high in fibre. On average, children between two and six years old in Florence eat less than 2 per cent of their diet as fibre. In contrast, the Boulpon children get over three times the proportion of fibre in their diets, at 6.5 per cent.

Taking a second look at the statistics for dietary intake in developed countries over the past several decades echoes that same difference. An adult in the UK in the 1940s consumed around 70 g of fibre per day, whereas now Brits average about 20 g per person per day.

We seem to be consuming far fewer vegetables. In 1942, we ate almost double the amount of veg that we eat now, and that was during the war when supplies were limited. Our intake of fresh green vegetables, such as broccoli and spinach, has suffered a precipitous decline which shows no sign of slowing. In the Forties, a typical daily serving of fresh green vegetables was about 70 g; in the last decade that has dropped to around 27 g. Our intake of beans, grains (including bread) and potatoes – all high in fibre – has also dropped since the 1940s. We simply eat fewer plant-based foods than we used to.

By looking at the genes contained within the Burkina Fasan microbiome, it's easy to see why the microbiota contains such a high proportion of *Prevotella* and *Xylanibacter* species – 75 per cent of the total species. Both of these bacterial groups harbour genes coding for enzymes that allow them to break down xylan and cellulose – two indigestible compounds that form the structure of plant cell walls. With *Prevotella* and *Xylanibacter* inside them, the children are able to extract far more nutrition from the grains, beans and vegetables that make up the bulk of their diets.

Italian children, on the other hand, have absolutely no *Prevotella* or *Xylanibacter*. They can't sustain these bacteria, because both varieties require an influx of plant remnants to survive. Instead, the Florentine kids' microbiotas are dominated by Firmicutes – the same group of bacteria that was found to be associated with obesity in several American studies. Indeed, the ratio of obesity-related Firmicutes to lean-associated Bacteroidetes in the Italian children was nearly three to one, compared with one to two in the Burkina Fasans.

A plant-rich diet, it seems, makes for a 'lean' set of gut microbes. So what happens if you give one group of Americans an animal-based diet of meats, eggs and cheeses, and another group a plant-based diet of grains, beans, fruits and vegetables? Not surprisingly, their gut microbes change composition. The plant-eaters saw a rapid increase in bacterial groups that break down plant cell walls, whereas the meat-eaters lost their plant-degrading bacteria and gained species that break down proteins, synthesise vitamins and detoxify the cancer-causing compounds found in charred meat. Their microbiomes began to resemble those of herbivorous animals on the one hand and carnivorous on the other. One volunteer in the study had been a lifelong vegetarian, but he was assigned to the animal-based diet group. His previously high *Prevotella* levels dropped as soon as he began eating meat, and within four days they had been outnumbered by microbes preferring animal protein.

This rapid adaptability goes to show just how useful teaming up with microbes can be when it comes to exploiting the food that's available at any given time. Using their adaptable microbes, our ancestors would have been able to make the most of a feast of fibre at harvest time, or a hunk of meat when they slaughtered an animal. It's a useful trick, especially for diets with unusual ingredients. Japanese people, for example, sometimes have microbes living in their guts with genes that code for enzymes that specifically break down the carbohydrates in seaweed. Because the seaweed *Porphyra* (known as *nori*) forms a big part of a sushi-based diet, one member of the typical Japanese microbiota, *Bacteroides plebeius*, has stolen genes coding for the porphyranase enzymes that digest *nori* from another species of bacterium, *Zobellia galactanivorans*, that lives on the seaweed. Probably, a great number of the microbes and microbial genes that allow us to exploit different foods originally came from bacteria living on those foods. Some suggest that our alliance with cattle has proved beneficial not only because of the extra meat and milk they give us, but also because of a transfer of the fibre-digesting microbes from their guts to ours.

The idea that the drop in our fibre intake might play a major role in the obesity epidemic does not mean that fat and sugar are irrelevant. High-fat, high-sugar diets are necessarily low in other macronutrients; namely, complex carbohydrates. Because most types of fibre *are* carbohydrates, including 'non-starch polysaccharides' such as cellulose and pectin, and 'resistant starches' found in foods like green bananas, whole grains, seeds, and even rice and peas that have been cooked and then cooled, increasing the relative fat and sugar content of the diet can mean the fibre content goes down. So could our diets be making us overweight, not because of their fat and sugar content, but because of the drop in fibre?

Remember Patrice Cani from chapter 2. He is the Professor of Nutrition and Metabolism at the Université catholique de Louvain in Belgium who discovered that lean people had much higher levels of the bacterium *Akkermansia muciniphila* than overweight people. He found that *Akkermansia* appeared to be encouraging the gut lining to strengthen, in part by forcing it to produce a thicker mucus layer. In doing so, this species was preventing the bacterial molecule LPS from getting across the gut lining and into the blood, where it caused inflammation in fat tissue, resulting in unhealthy weight gain.

Excited by the possibility that *Akkermansia* might help weight loss, or at least prevent gain, Cani tried feeding mice with a supplement of the bacteria. It worked. Not only did their LPS levels drop, but they lost weight. But adding *Akkermansia* is a temporary solution to a long-term problem. Without repeated top-ups, the population of the bacterium declines. So how can we keep its levels high after it has been added? And more pertinently for most of us, how can we boost the population of *Akkermansia* we may already have?

It was in trying to boost the population of another bacterial group – the bifidobacteria – that Cani found his answer. In feeding mice a high-fat diet, Cani had noticed that the abundance of bifidobacteria declined. In humans, too, it had been noticed that the higher someone's BMI, the fewer bifidobacteria they had. Cani knew that bifidobacteria were partial to a bit of fibre, and he wondered if providing a fibre supplement might boost their numbers in mice on a high-fat diet, and perhaps even halt weight gain. He tried supplementing their high-fat diets with a type of fibre known as oligofructose (sometimes called fructo-oligosaccharide, or FOS), which is found in foods including bananas, onions and asparagus. It made a big difference: the abundance of bifidobacteria rose.

But although bifidobacteria seemed to thrive on the oligofructose, it was *Akkermansia* that really took off. After five weeks, a group of genetically obese *ob/ob* mice whose diets were supplemented with oligofructose had eighty times the abundance of *Akkermansia* as *ob/ob* mice who didn't receive the fibre supplement. In genetically obese mice, adding fibre slowed weight gain, and in mice made fat by a high-fat diet, adding fibre induced weight loss.

Cani tried another type of fibre, called arabinoxylan, on his diet-induced obese mice. Arabinoxylan forms a huge part of the fibre found in whole grains including wheat and rye. Mice receiving a high-fat diet supplemented with arabinoxylan saw the same health benefits as those given oligofructose. It not only increased their bifidobacteria, but also restored their *Bacteroides* and *Prevotella* populations to the levels seen in lean mice. Despite the high-fat diets, giving fibre to these mice sealed up their leaky guts, encouraged their fat cells to become more numerous rather than grow larger, lowered their cholesterol levels, and reduced the speed of weight gain.

'If we feed mice a high-fibre diet at the same time as a high-fat diet, they can resist diet-induced obesity,' Cani says. 'We can argue that if we do not ingest a high fibre intake at the same time as a high fat intake, this will have a deleterious impact on the gut barrier. We know that our ancestors ate a diet very high in non-digestible carbohydrate. They were eating around 100 g of fibre a day, about ten times the amount we eat now.'

In our evolutionary past, weight gain would have been beneficial, allowing us to stock up in times of feast and survive through times of famine. Why weight gain appears to be so bad for us – inducing heart disease, diabetes and many cancers – is therefore puzzling. Many researchers say we have simply taken our weights so far beyond the human body's normal expectations for weight gain that we have crossed a line into weight-related ill-health. Cani's research provides a further clue as to why a process that was so beneficial throughout our evolutionary history is so detrimental now. Perhaps eating fat is not so bad, providing there's enough fibre in the diet to protect the gut lining from its effects. If fibre bolsters microbes that strengthen the gut wall's defences, LPS cannot get into the blood, the immune system can keep calm, and fat cells will become more numerous rather than filling up.

Actually, it may not be the microbes themselves that matter, but rather the compounds they produce when they break down dietary fibre. Namely, the short-chain fatty acids, or SCFAs that I mentioned in chapter 3. The three major SCFAs – acetate, propionate and butyrate – are present in the large intestine in substantial amounts after you've eaten plant foods. These products of the microbial digestion of fibre are the keys to a thousand locks, and their importance to our health has been underestimated for decades.

One such lock goes by the name of GPR43 (G-Protein-coupled Receptor 43). It belongs on immune cells, waiting for SCFA keys to come along and unlock it. But what does it do? As so often in biology, the easiest way to find out what something does is to break it, and watch what happens. By using GPR43 'knockout' mice that are missing their locks, a team of researchers discovered that without these

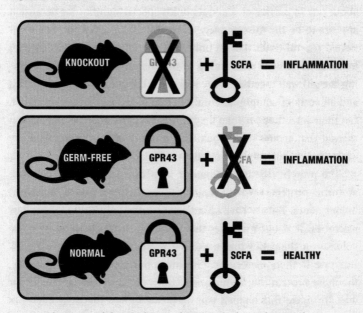

Schematic of the workings of GPR43 locks and SCFA keys.

receptors, mice get terrible inflammation and are prone to developing inflamed colons, arthritis or asthma. The same effect occurs if you leave the locks in place, but take away the keys. Germ-free mice cannot produce SCFAs, because they have no microbes to break down fibre. So their GPR43 locks remain closed and, again, the mice are prone to inflammatory diseases.

This intriguing result tells us that GPR43 is there to provide a route of communication between microbes and the immune system. By producing keys in the form of SCFAs, our fibre-loving microbes tap into the locks on immune cells, and tell them not to attack. GPR43 is not only found on immune cells, but also on fat cells. Here, when it is unlocked by an SCFA key, it forces fat cells to divide rather than grow larger, storing energy the healthy way. What's more, unlocking GPR43 with SCFAs also causes the release of leptin, the satiety hormone. In this way, eating fibre makes you feel full.

All three of the major SCFAs are important, but I want to tell you about one in particular. Butyrate holds special significance because it appears to be the missing piece in the puzzle of leaky gut. I've mentioned several times that an unhealthy community of microbes goes hand-in-hand with a loosening of the chains that hold the cells lining the gut wall together. Once loosened, the gut wall becomes leaky, and all sorts of compounds infiltrate the blood when they shouldn't. On their way, they provoke the immune system, and it is the inflammation that ensues that is behind a number of twenty-first-century illnesses. It is butyrate's job to plug the leaks.

The protein chains holding our intestinal cells together are, like any other proteins working behind the scenes in our bodies, produced by our genes. But we have ceded some control of those genes to our microbes. It is our microbes that decide how much to turn up the volume on the genes that make the protein chains of the gut wall. Butyrate is their messenger. The more butyrate they can produce, the more protein chains our genes churn out, and the tighter the gut wall. To make that happen you need two things: the right microbes (such as bifidobacteria to break down certain fibres into smaller molecules, and species like *Faecalibacterium prausnitzii*, *Roseburia intestinalis* and *Eubacterium rectale* to convert those smaller molecules into butyrate) and a diet that's high in fibre to feed them. They will do the rest.

The discoveries of Patrice Cani and others put a new spin on our understanding of the effect of the food we eat on weight. Rather than coming down to simply balancing calories-in with calories-out, the link between diet (and particularly fibre), microbes, SCFAs, gut permeability and chronic inflammation makes obesity seem rather more like a disease of energy regulation than a simple case of overeating. Cani believes poor diet is just one route to weight gain, and that anything that disrupts the microbiota, including antibiotics, can have the same effect if it means the gut allows LPS to pass into the blood.

Does this mean we can have our cake and eat it too, as long as it comes with a side of peas? Maybe. Study after study has shown that obesity is associated with low fibre intake. In one study of young American adults over the course of ten years, higher fibre intake was related to lower BMI, regardless of fat intake. In another that followed 75,000 female nurses over twelve years, those eating more fibre in the form of whole grains had BMIs consistently lower than those preferring refined grains that were low in fibre. Other studies have found that adding fibre to low-calorie diets boosts weight loss, with one showing that after six months on a 1,200-calorie/day diet, overweight women lost 5.8 kg with a placebo supplement, and 8 kg with a fibre supplement.

Altering your fibre intake over time also seems to have an effect on weight. Tracking the weights and fibre intakes of 250 American women for twenty months revealed that for every 1 g increase in their fibre intake per 1,000 calories, women lost 0.25 kg. It doesn't seem like much, but with a typical diet of 2,000 calories/day the women who increased their fibre consumption by 8 g/1,000 calories lost 2 kg. That's the equivalent of adding half a cup of wheat bran and half a cup of cooked peas to your daily diet.

Carbohydrates deserve special mention here. Like fats, and indeed fibres, not all carbs are created equal. Low-carb diet proponents claim all carbs are 'bad', but consider this: sugar is a carbohydrate, but so are lentils. Cake, for example, is around 60 per cent carbohydrate thanks to refined white flour and sugar, all of which is rapidly absorbed in the small intestine, but broccoli contains around the same amount, at roughly 70 per cent carbohydrate. Nearly half of this is fibre, consumed by microbes. The banner of 'carbohydrate' incorporates a huge spectrum of foods, from pure sugar through refined carbs such as white bread, to unrefined carbs like brown rice, which often have a high proportion of totally indigestible fibre. Low-carb diets give the impression that a spoonful of jam and a Brussels sprout are just as bad as one another, and as a result, these diets tend to be extremely low in fibre.

What carbohydrates do in your body depends greatly on exactly which kinds of molecules they contain. It doesn't just affect how many calories you absorb, but which microbes you encourage, and because of that, how your appetite is regulated, how much energy you store as fat, how permeable your gut is, how quickly stored energy is used, and the degree of inflammation in your cells. When it comes to carbohydrates, as Rachel Carmody points out, 'it really matters whether it's absorbed in the small intestine, or in the colon after being converted into SCFAs. And that is not captured on a nutrition label.'

Pulverising and juicing food also affects its fibre content. While 100 g of intact wholewheat grains contains 12.2 g of fibre, grinding it into wholewheat flour slightly reduces the fibre content to 10.7 g, and refining it into white flour drops it to just 3 g. A 250 ml fruit smoothie might contain 2 or 3 g of fibre, but the fruit that went into it would have provided 6 or 7 g if eaten intact. A 200 ml bottle of orange juice would have perhaps 1.5 g of fibre, whilst the four oranges that were juiced to make it contain eight times that amount – 12 g of fibre, pith and all.

Another food fad that likely impacts the microbiota is the raw-food diet. There is a school of thought, championed by Professor Richard Wrangham, a human evolutionary biologist at Harvard University – who was Dr Rachel Carmody's PhD adviser – that humans' use of fire to cook food prompted our transition as a species into being big-bodied, and even bigger-brained. As Carmody discovered in her PhD, cooking both animal and plant food changes the chemical structure and makes nutrients available to the body that were inaccessible when it was raw. The same is true of cooking's effect on the nutrients available to the microbiota. Not only that, but heat destroys some of plants' natural defensive chemicals that might otherwise kill beneficial microbes in the gut.

It's quite correct that eating raw food helps people to lose weight – there are simply fewer calories available to absorb. In fact, over the long term, the effect is so extreme that sustaining a healthy weight appears to be impossible on a pure raw-food diet. 'If you track rawfoodists over a long period of time,' Carmody says, 'you find that they can't maintain body mass. They're eating massive quantities of food, even in terms of calories, but they continue to lose weight. Strict rawfoodists can experience such severe energetic deficits that women of reproductive age actually stop ovulating.' From an evolutionary perspective, this is clearly not a great strategy. It indicates that cooking food is not merely a cultural invention, designed to improve the taste of food, but one that our species has physiologically adapted to, and that we are now obliged to continue. It's this impact of cooking our food on our gut microbiota that Carmody is now working to unravel.

If fibre is good, you might be wondering, then why are so many people intolerant to wheat and gluten? Wheat and other whole grains are packed with fibre and have many proven health benefits, from reducing the risk of heart disease and asthma, to improving blood pressure and helping to prevent strokes. Yet the popularity of the latest dietary fad – 'free-from' diets – is based on the idea that the gluten hiding away in wheat, rye and barley grains is bad for us.

Gluten is the protein in bread that gives it its spongy softness. Strands of gluten 'develop' as bread is kneaded, which then trap the carbon dioxide produced by the yeast so that the bread rises. It's a large molecule, a bit like a chain of pearls. The chain is partly broken by human enzymes in the small intestine, leaving smaller chains to go on to the large intestine.

Not so long ago, gluten-free foods, as well as lactose-free and casein-free, were completely absent from restaurants and super-markets, but the free-from market has blossomed over the past decade. It is no longer seen as a 'medical' diet reserved for people with unusual allergies and rare intolerances, but a lifestyle choice made by millions, and popularised by celebrities. The take-up of free-from foods follows a surge in food-blaming in recent years. Some people would have us believe that we are not 'meant' to eat wheat and that consuming dairy is not 'natural'. I even saw one website claiming that humans were the only species that drink the milk of other species, and therefore it was bad for us. No matter that this not-so-scientifically-grounded message came through the medium of the internet, also not known for its use amongst other species.

Wheat and dairy, and the compounds they contain – gluten, casein, lactose and so on – have been consumed by some human pop-ulations since the Neolithic Revolution, around 10,000 years ago, but they haven't caused problems on this scale until recently. This story takes us back to the Italian-born gastroenterologist Alessio Fasano, from the Massachusetts General Hospital for Children in Boston. He had started out attempting to create a vaccine for cholera, but found himself making the unexpected discovery of 'zonulin' – a protein that loosens the chains of the gut lining, making it leaky. He realised that zonulin was behind the autoimmune condition coeliac disease. Gluten was seeping through the gut walls of coeliac patients, which had somehow been made leaky by an excess of zonulin, triggering an autoimmune response that attacked the cells of the patients' guts.

Coeliac disease has become dramatically more common over the past few decades, and the only way to treat it is by avoiding even the tiniest amounts of gluten. But coeliacs are not the only ones shunning this protein. Millions of people consider themselves to be gluten in-tolerant, much to the delight of specialist food manufacturers and the consternation of many doctors. Proponents of a gluten-free diet claim that eliminating gluten not only reduces bloating and improves bowel function, but also gives them glowing skin, bags of energy and better concentration. Sufferers of irritable bowel syndrome are particularly keen on the diet. Lactose intolerance has also become more common, with supermarkets now stocking lactose-free products as standard.

But if wheat and dairy cause such trouble, why did our ancestors ever start consuming them? In the case of lactose – the sugar found in milk – people in many populations actually evolved what's known as 'lactase persistence'. We can all tolerate lactose as babies; it's in our mother's milk. We have a gene that makes an enzyme – lactase – specifically to break it down. Before the Neolithic Revolution, that gene would switch 'off' after infancy, when lactase was no longer needed. But during the Neolithic Revolution, some human populations started to herd animals: the ancestors of today's goats, sheep and cattle. At the same time, these populations began to evolve persistence of the lactase gene. No longer did it switch off after weaning, but instead remained on throughout adulthood.

Natural selection for lactase persistence happened *really quickly* in an evolutionary sense, which means one thing: if people are able to digest lactose as adults, it *really helps* them to survive and reproduce. In just a few thousand years, starting in the Near East, humans all across Europe shifted to lactose tolerance. Now, about 95 per cent of North and West Europeans can tolerate milk as adults. Elsewhere, other human populations who herd animals, like the goat-herding Bedouins of Egypt and the cattle-herding Tutsi of Rwanda, independently evolved lactase persistence through a different mutation than in Europeans.

The fact that so many of us are now unable to tolerate gluten and lactose is very unlikely to be evidence that we are not 'meant' to consume these foods. After all, the ancestors of many of us, especially those with European heritage, have been doing so for thousands of years. It's almost as if the lifestyle changes of the past sixty years have undone nearly 10,000 years of human dietary evolution. My point is not that dietary intolerances are not a real phenomenon, but that the origin of these conditions doesn't lie within our genomes, but rather within our damaged microbiomes. We *have* evolved to eat wheat and many of us, especially those of European ancestry, have evolved to tolerate lactose as adults, but we may have opened ourselves up to overreaction to these foods.

It is not the foods themselves, but what is happening to them within the body that is causing the problem. Unlike coeliac disease though, the gut lining is as intact as it should be in people with a sensitivity to gluten. Instead, it seems that through dysbiosis, the immune system has become unduly worried about gluten's presence. It probably doesn't help that the gluten content of wheat gets ever higher, in pursuit of the lightest, fluffiest bread, provoking an already irritable immune system. I'd like to think that rather than shunning gluten and lactose, we might manage to rekindle our post-Neolithic relationship with them at the same time as we restore our microbial balance.

The American food writer Michael Pollan famously said that we should 'eat food, not too much, mostly plants'. Though he wrote it before the revolution in our understanding of the microbiota, we now know it to be truer than ever before. By avoiding food that's packaged in cardboard boxes, kept 'fresh' through chemical preservatives with questionable long-term safety profiles; by not filling ourselves beyond the point that our pancreases, adipose tissue and appetites can keep up; by remembering that plants feed both us and our microbes, we can nurture a microbial balance that underpins good health and happiness.

I have made the case for fibre in this chapter, but it's worth emphasising that no element of our diets stands alone. The great complexity of the pros and cons of any food type, whether it's the various forms of fats or the size of carbohydrate molecules, must fit within the framework of the diet as a whole. It's not enough to say that fats are bad and fibre is good, because the old adage – everything in moderation – holds true no matter what the latest dietary fad proclaims. Fibre's value is in its effects on the particular community of microbes that our species has cultivated on our journey from herbivorous beginnings to an omnivorous present. The anatomy of our digestive system, with its emphasis on the large intestine as a home for plant-loving microbes, and a long appendix that acts as a safe-house and storage facility, serves to remind us that we are not pure carnivores, and plants are our staple diet. The nutrient we're missing out on is fibre, but it is *plants* that we're forgetting to eat.

I sometimes think how lucky we are to have a physiological obligation to eat every day. It's one of life's greatest pleasures, and it is *essential*. Barely any other human activity is both as enjoyable and as necessary for survival. But there has to be a balance between these two aspects of eating: the hedonism and the sustenance. The irony is, that those of us living in developed countries have access to the most plentiful, fresh, varied and nutritious food on Earth, whatever the season, and yet a vast number of us will die from diet-induced and diet-related diseases – more than die from mal- and under-nutrition. Yes, we can lay plenty of blame at the doors of multinational food companies that fill their products with sugar, salt, fats and preservatives. Of course, we need to know more about the consequences of intensive, medicalised farming practices. Undoubtedly, it's true that doctors and scientists don't have all the answers when it comes to the perfect balance of nutrients. But ultimately, we each hold the responsibility for our own diets, and those of our children, and we have the freedom to take control of what we eat.

You are what you eat. What's more, you are what *they* eat. With each meal you make, spare a thought for your microbes. What would *they* like you to put in your mouth today?

SEVEN

From the Very First Breath

At six months old, a koala joey begins to peep out from its mother's pouch. It's time to start making the transition from feeding solely on milk to its adult diet of eucalyptus leaves. To most herbivores, it's not the most appetising of diets; the leaves are tough, toxic and almost devoid of nutrients. The mammalian genome is not even equipped with the genes necessary to produce enzymes that can extract anything of worth from eucalyptus. But koalas have found a way around this problem. Like cows, sheep and many others, koalas make use of microbes to extract the bulk of the energy and nutrients they need from fibrous plant material.

The trouble is, joeys do not have the microbes they need to break down eucalyptus leaves. It's up to their mums to sow the seeds of a microbial community in their guts. When the time is right, mother koalas produce a soft, runny substance called 'pap' – a faeces-like paste made up of predigested eucalyptus and an inoculum of gut bacteria. They feed this pap to their joeys, providing them with not only the beginnings of a microbiota, but enough microbe-food to get the colony started. Once it has taken hold in the gut, a joey has its own tiny workforce that transforms its digestive abilities and makes eucalyptus edible.

Getting your microbiota from your mum is common, even among non-mammals. Cockroach mums keep their microbiotas in specialised cells called bacteriocytes, which expel their microbial contents alongside a developing egg inside her body. The egg then engulfs the bacteria before it is laid. Stinkbug mothers, on the other hand, take a more koala-like approach to endowing their offspring with useful microbes, smearing the surface of their eggs with bacteria-laden faeces as they deposit them. After the eggs hatch, the nymphs immediately consume the faecal smear. Another species, the Kudzu bug, hatches without any microbes, then consumes a bacteria-filled package left by its mother near to its egg. If the package is missing, the bugs show bizarre wandering behaviours, searching for packages belonging to other eggs. Birds, fish, reptiles and others are also known to pass their microbiota on to their offspring, either inside the egg or as the young are born.

Whatever a species' parental proclivities, providing one's offspring with a good set of microbes to help them on their way seems to be a near-universal ritual. That it is so common speaks volumes for the evolutionary benefits of a life lived in the company of microbes. If egg-smearing behaviour and bacteria-engulfing mechanisms are the norm, they have evolved to be so. They must be enhancing the survival, and reproductive abilities, of the individuals that possess them. So what of us humans? It's clear that our microbiotas benefit us, but how do we ensure that our little ones receive the seeds of their own colonies?

In the first few hours of a baby's life, it goes from being mostly human to mostly microbial, in terms of cell numbers at least. Bathing in its warm sac of amniotic fluid inside the uterus, the baby is protected from the microbes of the outside world, including those of its mother. But once the waters break, colonisation begins. The baby's journey out of its mother is a microbial gauntlet. In fact, 'gauntlet' gives the wrong impression, because the microbes the baby encounters are not enemies, but friends. Rather, it is a microbial rite of passage, coating the previously near-sterile baby in a smear of vaginal microbes.

As it emerges from its mother, it gets another dose of microbes alongside those from the vagina. Disgusting as it may sound, ingesting faeces early in life is not unique to koalas. During human labour and birth, the contraction-inducing hormones and the pressure of the descending baby cause most women to defecate. Babies tend to be born head first and facing towards their mum's bottom, pausing for a moment with their heads and mouths in prime position whilst their labouring mothers wait for the next contraction to help them ease the rest of the body out. Whatever your instinctive revulsion, it's an auspicious start. After birth, the mother's gift of a new coat of microbes, both faecal and vaginal, makes for a simple and safe birthday suit for the newborn.

It's also probably an 'adaptive' start. That is, it's probably no bad thing that the anus is so close to the vagina, or that the hormones that bring on contractions in the uterus have the same effect on the back passage. Natural selection may have made it that way because it benefits baby, or at the very least causes no more harm than good. Receiving a gift of the microbes, and their genes, that have worked in harmony with your mother's genome gets you off to a great start.

In a comparison of the gut microbes of a young baby with samples taken from the mother's vagina, stool and skin, and the father's skin, it is the strains and species of the mother's vagina that are most similar to those colonising the newborn's gut. Species belonging to genera like *Lactobacillus* and *Prevotella* are most common. These vaginal microbes are quite a select bunch – far less diverse than those in the mother's gut – but they seem to have a specialist role in the newborn's developing digestive tract. Where there are *Lactobacillus*, the thinking goes, there are no pathogens. No *C. diff*, no *Pseudomonas*, no *Streptococcus*. These nasties can't get a foothold, because the lactobacilli (that's just *Lactobacillus* species en masse) crowd them out. They are part of a group known as the lactic acid bacteria, which includes the species that convert milk into yogurt. Not only does the lactic acid (which gives yogurt its sour taste) create an environment hostile to other bacteria, but lactobacilli also produce antibiotics of their own. Called bacteriocins, lactobacilli produce these chemicals to kill off pathogens rivalling them for prime real estate in the empty gut of the newborn.

But why is it that the microbes colonising babies' guts are more similar to those in their mother's birth canal than in her gut? If gut microbes are there to help us digest food, wouldn't they be more suitable? Doctors, and many women, are well aware that lactobacilli thrive in the vagina – a dose of live yogurt has long been a home remedy for a bout of thrush (a yeast infection). It's often assumed that these lactic acid bacteria are there to protect the vagina from infections, but although they do a good job of this, it's not their primary purpose.

The vagina's lactic acid bacteria are milk-eaters. They take the sugar found in milk – lactose – and convert it into lactic acid, creating energy for themselves in the process. Babies are also milk-eaters, converting lactose into two simpler molecules, glucose and galactose, which are absorbed by the small intestine into the blood and then converted into energy for the baby. Any lactose that makes it past the small intestine undigested is not wasted. It goes straight to the lactic acid bacteria waiting in the large intestine. The lactobacilli that a baby acquires as it passes through the birth canal, then, are not there to protect mum's vagina, they are there to colonise baby. Perhaps it seems over the top to have a vagina permanently colonised by lactic acid bacteria, but giving birth is what it's there for, and with far greater frequency than the women of the developed world do now. The vagina is a baby's starting gate, and it has evolved to provide the best beginning for the race of life.

Although babies benefit from lactic acid bacteria in the early days of their lives, ultimately they will need a gut microbiota that does more than break down milk. They need some of their mother's gut microbes. Aside from a faecal greeting at birth, they actually acquire some gut microbes from their mother's vaginal microbiota too. The bacterial communities living in pregnant women's vaginas are different ones from those living in the vaginas of non-pregnant women. Among the normal species of the vagina are species more normally found in the gut.

Take *Lactobacillus johnsonii*. It's usually found in the small intestine, where it produces enzymes that break down bile. But during pregnancy, its abundance in the vagina shoots up. It is an aggressive little thing, producing lots of bacteriocins that can kill off threatening bacteria, giving it more room to itself and thus a better showing in the baby's gut.

During pregnancy the vaginal microbiota shifts, growing less diverse. It's almost as if it's paring down the community in preparation for seeding baby with its first, and most essential, microbes. When a baby is colonised, its gut microbes are relatively diverse, including some of mum's faecal (gut) bacteria as well as the vaginal set. But this initial collection also rapidly slims down to just those bacteria that can help with milk digestion. It's my guess that the faecal species baby gets from mum might be quickly squirrelled away in the safe-house of the appendix, ready for later use.

The early colony that takes up residence in a baby's gut forms a crucial starting point for the microbiota that will develop over the next few months, possibly setting the trajectory for several years. In our macro-scale world, a bare patch of rock will accumulate lichens, then mosses, until the turnover of these pioneers results in enough soil to sustain small plants, and eventually bushes and trees. A bare patch of ground might one day become an oak woodland in Britain, a beech–maple forest in the United States, or a tropical rainforest in Malaysia. The same goes for the 'bare ground' of the gut – the microbiota begins with a simple selection of lactic acid bacteria, then grows increasingly complex and diverse. This is ecological succession – each stage provides the habitat and the nutrients necessary for the next.

Micro-scale succession takes place in the baby's gut and across its skin, too. The initial colonisers – the pioneers – influence which species settle there next. Just as an oak forest provides acorns and a tropical rainforest provides fruits, so the microbiota produces different resources for the growing baby. These play a role in refining the baby's metabolism and educating its immune system. Immune cells, tissues and vessels grow and develop under the instruction of a self-interested, yet benevolent, microbiota. Thanks to a healthy dose of vaginal microbes, the baby's microbial partnership is under way, and off to a good start.

So far, so good. Except that millions of babies are born each year without going anywhere near their mothers' vaginas. In some places, giving birth by Caesarean section is more common than giving birth vaginally. Nearly half of women in Brazil and China have their babies cut from their bellies rather than giving birth. Taking into account the number of women living in rural areas with no access to a hospital in these countries, the average rate of C-sections in cities is likely to be even higher. Indeed, in some hospitals in Rio de Janeiro, the rate of C-sections exceeds 95 per cent of births. The shocking case of Adelir Carmen Lemos de Goés in 2014 gives some idea of just how entrenched C-sections are in Brazilian society. Having already had two Caesareans, Adelir wanted to give birth vaginally, and checked herself out of hospital when she was told that wouldn't be possible. She made her way home to have her baby, but a short time later was picked up by armed police, taken back to hospital, and forced to give birth by Caesarean. It seems vaginal deliveries are considered too time-consuming and unpredictable for many Brazilian health-care facilities.

Even in countries where a woman's choice is respected, C-sections are surprisingly common. Many women are told that it's once a C-section, always a C-section, because the scar in the uterus from previous surgery could rupture under the pressure of contractions, but this is not true. It's taking time to filter down from researchers to those setting medical guidelines and on to medical staff, but there's now thought to be no significant extra risk to giving birth vaginally after up to four previous Caesareans. In some hospitals in the United States, up to 70 per cent of births, and possibly even more than that, are by C-section. The average across the country is a substantial 32 per cent. It's pretty typical for between a quarter and a third of births to be by C-section in developed countries, and many developing countries aren't far behind. In fact, quite a number are steaming ahead: the Dominican Republic, Iran, Argentina, Mexico and Cuba are all in the high 30s and 40s.

Needless to say, it hasn't always been this way. C-sections have been used sparingly for centuries, usually to save a trapped baby from being pulled under by its dying mother. The last century, though, brought tolerable anaesthetics and improved surgical techniques, and with these came the opportunity to save not just babies, but their mothers too. Where once a baby might be starved of oxygen in a difficult delivery, or a mother might haemorrhage, C-sections provided a safer alternative. From the late 1940s onward, they grew more and more common as antibiotics took the edge off the major risk to the mother. The 1970s saw a sharp upturn, and the rise has continued nearly unabated since. C-sections are now the most common abdominal surgery undertaken.

Much of the popular press, fuelling 'mommy wars' at every opportunity, would have you believe that this rise is because ever-growing numbers of women are 'too posh to push', opting for a quick, convenient and pain-free alternative to hours of labour, as well as a vagina-sparing birth. Although rates of pre-labour elective C-section are growing, a greater part of the rise is actually due to extra Caesareans performed *during* labour, following recommendation by midwives and obstetricians. Americans are fond of blaming the sue-happy culture of privatised health care for the risk-averse behaviour of medical staff when it comes to a challenging labour or birth. But even within publicly run health services such as the UK's National Health Service, doctors are turning more readily to C-sections when the going gets tough. We're talking about labours that aren't moving fast enough, large babies that might get stuck, or breech births, where the baby is heading out bottom-first and can't be turned around.

For many women, the medical recommendation of a C-section during the labour process comes as a guilty relief. It offers an opportunity to escape from the pain, exhaustion and fear of a vaginal delivery. The very existence of a readily available alternative to a natural birth disempowers women, making them feel that pushing their baby out is too difficult or dangerous to risk. The reality is otherwise. For mothers, on average, the risks involved with having a planned C-section actually exceed the risks of a vaginal delivery. In France, for example, around four previously healthy women die in every 100,000 giving birth vaginally, but thirteen or so die after a C-section. Even in non-fatal circumstances, C-sections are more dangerous than vaginal deliveries: infections, haemorrhaging and problems with anaesthesia can occur – all risks of any abdominal surgery.

C-sections are a crucial alternative to vaginal delivery in medically necessary circumstances – some women have no choice but to give birth this way. The World Health Organisation estimates that the optimal rate of C-sections should lie between 10 and 15 per cent of all births. It's at this point that women and their babies are safe from the dangers of childbirth, but not exposed to any unnecessary risks of surgery. For doctors, knowing *which* 10 or 15 per cent of the women they should perform C-sections upon presents a challenge. For women choosing to have an elective pre-labour C-section, the risks to them and their baby are often not fully explained, or even appreciated by their carers.

As it stands, the major risks of C-section to the baby tend to relate to its first few days and weeks of life. Let's hear from the UK's National Health Service on the risks of C-section:

Sometimes a baby's skin may be cut when the opening in the womb is made. This happens in 2 out of every 100 babies delivered by caesarean section, but usually heals without any further harm. The most common problem affecting babies born by caesarean section is difficulty breathing, although this is mainly an issue for babies born prematurely. For babies born at or after 39 weeks by caesarean section, this breathing risk is reduced significantly to a level similar to that associated with vaginal delivery. Straight after the birth, and in the first few days of life, your baby may breathe abnormally fast. Most new borns recover completely within two or three days.

But there is rarely mention of the longer-term impacts of birth by C-section on the baby as it ages. What was once assumed to be a harmless alternative to vaginal delivery is increasingly recognised as carrying risks to the health of both mother and baby. In the early days, for example, Caesarean babies are more susceptible to infections. Up to 80 per cent of cases of MRSA infections in newborns occur in those born by Caesarean. As toddlers, children born by C-section are more likely to develop allergies. Those born to allergic mothers – who are probably predisposed to allergy – are seven times as likely to be allergic themselves if they were delivered by C-section.

C-section babies are also more likely to be diagnosed autistic. If no babies were born by C-section, researchers at the Centers for Disease Control (CDC) in the United States estimate that 8 out of every 100 autistic children would be protected from developing the disease. Similarly, people with obsessive–compulsive disorder are twice as likely to have been born by C-section. Some autoimmune diseases have also been associated with Caesarean delivery. Both type 1 diabetes and coeliac disease are more likely in children born by C-section. Even obesity is associated with having been born by C-section. In a study of young adult Brazilians, 15 per cent of those born by C-section were obese compared with 10 per cent of those born vaginally.

You've probably noticed the connection here. These are twenty-first-century illnesses. Though each one is multi-factorial, stemming from a wide range of environmental risk factors and genetic predispositions, the overlap between Caesarean births and an increased risk of twenty-first-century illnesses is striking. Sampling the gut microbiota of babies can reveal whether they were born by C-section or vaginally for many months after their birth. The vaginal microbiota that colonises a newborn's body, both inside and out, as it emerges from the birth canal cannot colonise a baby delivered out of the sunroof. Instead, it is the microbes of its environment that a C-section baby first encounters, as its little body is pulled by gloved hands past the skin of its mother's belly, shown to its anxious parents, then whisked across the operating theatre to be towelled down and checked over. In a sterile surgery, that might mean the hardiest of hospital bugs – *Streptococcus*, *Pseudomonas* and *Clostridium difficile* perhaps – as well as the skin of mother, father and medical staff. It is these skin microbes that form the basis of a C-section baby's gut microbiota.

But whereas the microbiotas of mother's vagina and the child's gut match following a natural birth, C-section babies and their mums cannot be matched up by looking at their microbes. In place of lactose-digesting bacteria – *Lactobacillus*, *Prevotella* and the like – that vaginal delivery provides are the stalwarts of the skin: *Staphylococcus*, *Corynebacterium*, *Propionibacterium* and so on. These are not lactose-digesting bacteria, they are sebum- and mucus-loving species. What should be the foundation for an oak woodland is instead the beginnings of a pine forest.

How exactly this difference in gut microbiota translates to any of the health consequences associated with Caesarean birth is steadily emerging through research. Better knowledge of these mechanisms would take this microbial connection from being a serious concern to a certain consequence. But whatever the workings behind it, worries about the impact of delivery mode on the future development of the gut microbiota were enough to prompt microbiome scientist Rob Knight into action when his wife had to give birth to their daughter by emergency C-section in 2012. Having been involved in several research studies into the development of the infant gut microbiota at the University of Colorado in Boulder, Knight was keen to try and prevent any negative consequences for his newborn daughter of missing out on a vaginal delivery. After waiting for the medical staff to leave the room, he used a swab to transfer vaginal microbiota from his wife to his daughter.

While this subversive act may not have garnered much support amongst most medical staff on the delivery ward, it holds great potential. Rob Knight and Maria Gloria Dominguez-Bello, an associate professor in the Department of Medicine at New York University, are now running a large clinical trial to establish whether transferring microbes from a woman's vagina to her newborn might ameliorate some of the short- and long-term effects of C-sections. The experimental technique is simple: a small piece of gauze is inserted into the mother's vagina an hour before she is due to go into the operating theatre. Just before the first cut is made, the gauze is removed and placed in a sterile pot. A few minutes later, when the baby is out, it is rubbed with the gauze – first in its mouth, then across its face, and then over the remainder of its body.

It's a simple but effective intervention. Preliminary results from seventeen babies born to women in Puerto Rican hospitals show that the inoculated babies had gut microbiotas that were far closer to their mother's vaginal and anal microbiotas in comparison with babies born by C-section but not swabbed. Although swabbing didn't completely normalise these babies' microbiotas, the impact was significant, boosting numbers of species normally seen in babies born vaginally.

The microbial consequences of vaginal and Caesarean births raise some intriguing questions that we just don't have answers to yet. What, for example, is the impact of a water birth on baby's first inoculum? What does the warm water, possibly infused with the remnants of an antibacterial cleaner used on the bath's surface, do to the vaginal microbiota and its transfer to a baby's skin and mouth? And what of caul births, where the newborn arrives still cloaked in the amniotic sac, missing contact with its mother's genital microbes altogether? And how, microbially speaking, do home births compare with those in the supposedly cleaner environment of a hospital?

Even vaginal births are relatively germ-free affairs in the Western world. In comparison with giving birth, often at home, in large areas of Africa, Asia and South America, giving birth in much of Europe, North America and Australasia is a highly medicalised, sterile process. Beds, hands and tools are all washed with antibacterial soaps and alcohol rubs before they come into contact with the labouring woman or her baby. Nearly half of American women are put on an antibiotic drip to prevent them from passing on harmful bacteria such as Group B strep to their babies. And *all* American babies receive a dose of antibiotics straight after they are born, just in case their mothers have gonorrhoea, which could, in rare cases, cause an eye infection. Ignaz Semmelweis would be pleased to see his antiseptic measures so thoroughly and effectively put into practice, and there's no doubt that many thousands of mothers and babies are alive because of such hygiene. But it is *different* from what the human genome and the microbiome expect. It's this difference and its consequences that should guide the next step in improving medical care for women and babies.

Ultimately, this is not an issue for women to wrestle with or feel guilty about alone. It isn't the relatively small subset of women who choose a pre-labour elective C-section who need to change, it's the whole culture of the medicalisation of birth. Many initiatives already exist to reduce the rate of Caesarean sections around the world, most of which focus on its risks for the mother, and the use of scarce resources on unnecessary surgical procedures. Added to these concerns should be a greater understanding of the hazards, and the benefits, of Caesarean deliveries for the health of newborns, both in the short and long term.

With the first few seconds of life, and the microbes that accompany them, now elapsed, the seeds of a baby's microbiota still have a long way to go to reach maturity. What grows next depends on how those seeds are tended in the days, weeks and months to come.

In 1983, Professor Jennie Brand-Miller became a mother. A few days later she was forced to develop a particular interest in infantile colic. Her baby son was crying inconsolably, despite being apparently healthy. Brand-Miller and her husband had both previously been involved in research on lactose intolerance – the inability of some people to break down the milk sugar lactose using lactase enzyme. They wondered whether this might be the cause of the colic. Brand-Miller's husband focused on answering this question in his PhD, organising a placebo-controlled trial of lactase enzyme droplets for colicky babies. Unfortunately, there was no difference in the babies' time spent crying between the lactase and the placebo. But there *was* a difference in the amount of hydrogen in the breath of colicky and non-colicky babies.

This got them thinking. Excess hydrogen in the breath indicates that bacteria in the gut are breaking down food. Lactose, though, should mostly be broken down by the body's own enzymes into two smaller sugars: glucose and galactose. If bacteria were producing hydrogen, they must have been receiving a sizeable meal; suggesting that some other molecule that was not broken down in the small intestine was getting through. Brand-Miller knew of a set of compounds called oligosaccharides that formed a large part of breast-milk. But these molecules were thought to be pointless, as the human body does not possess the necessary enzymes to break them down. She and her husband had a hunch. Perhaps the oligosaccharides were not there to feed the baby, but to feed its gut bacteria?

Oligosaccharides are carbohydrates made up of short chains of simple sugars (oligo means 'few'). Human breast-milk contains an enormous variation of them – around 130 different kinds in all. This is far more than the milk of any other species; cow's milk, for example, has just a handful of varieties. As adults, we don't eat anything containing these molecules. Yet they are manufactured in the breast tissue of pregnant and breast-feeding women. This is a clue as to their importance – why specifically manufacture something if it has no function?

To test their theory, Brand-Miller and her husband ran a trial. They measured the amount of hydrogen in babies' breath when they were given glucose in water, and also when they were given purified oligosaccharides in water. There was no increase in hydrogen with the glucose, indicating that it was absorbed in the small intestine and not broken down by the gut bacteria. But the oligosaccharide mix produced a spike in hydrogen levels – these compounds were going straight through the small intestine and feeding the gut microbiota, not the baby.

Oligosaccharides are now known to be instrumental in encouraging the right species of microbes to bloom in the seedling gut microbiota of a baby. Babies fed breast-milk have microbiotas dominated by lactobacilli and bifidobacteria. Unlike the human body, bifidobacteria make enzymes that can use oligosaccharides as their sole food source. As a waste product, they produce the all-important short-chain fatty acids (SCFAs) – butyrate, acetate and propionate, plus a fourth SCFA that's particularly valuable in babies: lactate (also known as lactic acid). These feed the cells of the large intestine, and play a crucial role in the development of a baby's immune system. Simply put, where adults need dietary fibre from plants, babies need oligosaccharides from breast-milk.

Acting as a food source for bacteria is not the only function of the oligosaccharides found in human breast-milk. In the early days and weeks of life, a baby's gut microbiota is both very simple and highly unstable. Strains of bacteria go through booms and busts. This leaves the community vulnerable to disaster. The entry of one pathogenic strain – *Streptococcus pneumoniae* for example – can wreak havoc, decimating beneficial strains. Oligosaccharides provide a mopping-up service. Before a pathogenic bacterium can do any damage, it must adhere to the intestinal wall using special attachment points on the bacterial surface. Oligosaccharides fit perfectly into these points, depriving nasty species of a foothold. Of the 130 or so compounds, dozens are known to be specific to particular pathogens, fitting their attachment sites like a lock and a key.

The composition of breast-milk adapts to the growing baby's needs as it ages. Just after birth, the initial milk, called colostrum, is thick with immune cells, antibodies and a good four teaspoons' worth of oligosaccharides in every litre of milk. Over time, as the microbiota stabilise, the oligosaccharide content of the milk decreases. By four months after birth it has dropped to less than three teaspoons' worth per litre, and by the baby's first birthday it contains less than one.

Once again, we can take instruction from the koala, and other marsupials, this time about the importance of the oligosaccharide content of milk. Most marsupials have two teats, which are inside the pouch. Only one of these is used by the joey for the duration of its suckling life. If two joeys are born in two consecutive seasons, they each have their own teat. Remarkably, the two teats provide milk that is tailored to the age of each joey. The newborn receives milk high in oligosaccharides and low in lactose, whereas the older joey receives a lower dose of oligosaccharides, but far more lactose. Once a joey has left the pouch, the oligosaccharide content of its milk supply drops even further.

This custom-made milk shows that oligosaccharides are not declining simply because the mother cannot keep up their production. Rather, marsupials have adapted to provide their offspring with milk appropriate to their changing microbial communities. Nature has selected for milk that benefits microbes, because microbes benefit mammals.

Oligosaccharides aren't the only surprise ingredient in breast-milk. For decades, kind-hearted breast-feeding mums have donated their milk to hospital milk banks. They provide a lifeline to babies whose mothers can't breast-feed them, often because the baby was premature or too ill to start breast-feeding, so the mother's supply has dried up. But milk banks have an enduring problem: donated milk is always contaminated with bacteria. Many of these microbes come from the skin of the nipple and breast, but however stringently a donor's skin is sterilised before collection, microbes in breast-milk are never eliminated.

Increasingly sophisticated aseptic milk-collection techniques, combined with DNA-sequencing technology, have revealed the reason behind this so-called contamination. The bacteria *belong* in the breast-milk. They are not hitchhikers from the mouth of the baby or the nipple. Instead, they have been packaged up in the breast tissue itself. But where have they come from? Many are not typical skin-dwelling bacteria that have sidled into the milk ducts from their usual home on the skin of the breast. Instead, they are lactic acid bacteria, more normally found in the vagina and the gut; indeed looking in the mother's stool reveals matching strains in her gut and her milk. Somehow, these microbes have travelled from the large intestine to the breasts.

Checking the blood for migrant microbes shows the route they took. They are stowaways, travelling inside immune cells called dendritic cells. The dendritic cells are willing participants in the trafficking of the bacteria. Sitting among the dense immune tissue that surrounds the gut, these cells are able to reach out with long arms (dendrites) into the intestine to check what microbes are present. Usually, they are responsible for engulfing pathogens, then waiting for another team of immune cells – the 'natural killer cells' – to turn up and destroy them. Extraordinarily, the dendritic cells can also pluck unsuspecting beneficial bacteria from among the crowd to engulf and transport through the blood to the breasts.

It's possible to see the system at work in a study of mice, too. Whereas only 10 per cent of non-pregnant mice had bacteria in their lymph nodes, 70 per cent of heavily pregnant mice did. After giving birth to their pups, the numbers of bacteria in the lymph nodes dropped sharply, but at the same time the proportion of mice with bacteria in their mammary tissue shot up to 80 per cent. In both mice and humans, it seems the immune system is not only working to keep the bad bugs out, but to bring the good bugs in so that they can be passed on to the newborn baby. It's a great strategy: the bacteria gain a new home with few competitors for the space, and the baby gains a supply of useful bacteria to supplement those it received during birth.

Just as the oligosaccharide content of breast-milk changes as the baby ages, so does the mix of microbes the milk contains. The species a baby needs on day one are different from those it needs at one, two and six months old. The colostrum produced in the first few days after the baby's birth contains hundreds of species. Genera such as *Lactobacillus*, *Streptococcus*, *Enterococcus* and *Staphylococcus* have all been detected in breast-milk, in quantities up to 1,000 individuals per millilitre of milk. That means a baby might consume around 800,000 bacteria each day from its milk alone. Over time, the microbes in milk grow less numerous, and shift towards different species. More of the types of microbes that are found in an adult mouth are present in the milk produced several months after birth; perhaps preparing the baby for its introduction to solid food.

Curiously, how a baby is delivered has a major impact on the microbes found in its mother's breast-milk. The microbial content of the colostrum of women who delivered by pre-labour elective Caesarean section is strikingly different from that of women who delivered vaginally. This difference persists until at least six months. But in women who had in-labour emergency C-sections, the milk microbiota was far more similar to that of mums who delivered vaginally. Something about the labour process sounds a klaxon, informing the immune system that it's time to prepare for the baby to be on the outside of the body, receiving nourishment from breast-milk instead of the placenta. It seems likely that this klaxon comes in the form of the many powerful hormones released during labour. Their release alters which microbes are moved from the gut to the breasts in preparation for the emerging baby. C-section therefore creates a double whammy of difference: not only does it alter the microbial inoculum a baby receives on entering the world, but the follow-up microbes it receives in the breast-milk.

The oligosaccharides, live bacteria and other compounds found in breast-milk make for an ideal food for both babies and their microbiotas. Breast-milk encourages the settling of beneficial microbes and guides the gut microbiota towards an adult-like community. It prevents colonisation by harmful species, and educates the naive immune system as to what is worth worrying about, and what deserves to stay.

So what of bottle-feeding (by which I mean infant formula feeding)? How does formula milk affect a baby's developing microbiota? Fashions in infant feeding are as capricious as those in skirt lengths. Even before bottle-feeding had become a realistic option for mothers, there was an alternative to breast-feeding one's own child. Wet nursing was common before the twentieth century, with trends amongst the social classes shifting in a similar way to those in bottle-feeding over the past century. At one moment it might be deemed unseemly for aristocratic women to feed their own babies, and at another the working women of the Industrial Revolution were taking on wet nurses while the social elite returned to nursing their own babies.

In the late nineteenth and early twentieth centuries, wet nurses lost their jobs to increasingly practical bottle-feeding alternatives. Easy-to-sterilise glass bottles, washable rubber teats and modified cow's milk formulas took bottle-feeding from an alternative of need to an alternative of choice. Breast-feeding rates plummeted. In 1913, 70 per cent of women breast-fed their own newborn, but this figure dropped to 50 per cent in 1928 and just 25 per cent by the end of the Second World War. In 1972, breast-feeding reached an all-time low of just 22 per cent. After tens of millions of years of mammalian lactation, feeding breast-milk to newborn babies all but ground to a halt within a single century.

If the oligosaccharides and live bacteria in breast-milk are responsible for nourishing the seeds of a baby's gut microbiota, changing in sync as it grows, what are the microbial consequences of bottle-feeding? Bottled milk is, more often than not, still 'breast' milk, but from the breast of a cow, not a human. Cow's milk has evolved, despite a lot of meddling by humans over the past 10,000 years, to be the ideal nourishment for calves and their microbes. But the gut microbiota of a calf is a dramatically different one from that of a human child. It thrives on doubly-chewed grasses, not the remnants of meat and veg predigested in the small intestine. Cow's milk alone leaves a lot to be desired as a newborn baby food, often leaving infants with vitamin and mineral deficiencies that may cause scurvy, rickets and anaemia. Modern infant formulas are supplemented with many essential extras, but they don't usually contain immune cells and antibodies, oligosaccharides or live bacteria.

The most obvious difference in the gut microbiotas of bottle-fed babies is the sheer diversity of species and strains. Babies that are not breast-fed at all have around 50 per cent *more* species living in their guts. In particular, exclusively bottle-fed babies had far more species from the group Peptostreptococcaceae, which includes the nasty pathogen *Clostridium difficile*. If *C. diff* takes over, it can cause intractable diarrhoea and is fatal in children frighteningly often. Whereas a fifth of babies who are exclusively breast-fed carry *C. diff*, nearly four-fifths of babies who are put straight onto formula carry the bug. It's likely many of these children picked it up in the delivery room – the longer a hospital stay, the more likely a newborn is to pick up this pathogen.

Whereas high diversity of microbes seems to be an indicator of good health in adults, in babies the opposite is true. Cultivating a very select group of species in the early days of life, with the aid of the vagina's lactic acid bacteria and breast-milk's oligosaccharides, appears to be important in protecting babies from infection, and priming their young immune systems. Even combining breast-feeding with bottle-feeding increases the unwanted diversity of microbes, including *C. diff*, producing a microbiota with a composition part way between those of exclusively breast-fed and exclusively bottle-fed babies.

But does it really matter if babies have a little more diversity in their bellies? Can encouraging a different group of bacteria really do any harm? The fact that breast is best is often recited, but with little appreciation for what that actually means for a child's health. It implies that formula-feeding is good, and breast-milk is a bonus. But looking at the data shows a stark contrast between the health of breast-fed and bottle-fed babies.

To start with, bottle-fed babies are more prone to infections. When compared with solely breast-fed babies, exclusively bottle-fed babies are twice as likely to get ear infections, four times as likely to end up in hospital with a respiratory tract infection, three times as likely to suffer from a gastrointestinal infection, and two and a half times as likely to get necrotising enterocolitis, in which the bowel tissue dies. They are also up to twice as likely to die from Sudden Infant Death Syndrome. In the United States, infant mortality (death before the age of one year) is 30 per cent higher among babies who were not breast-fed, even accounting for numerous other factors, such as smoking in pregnancy, poverty and education, and excluding babies whose illness prevented them from being breast-fed. Infant mortality is already low in developed countries, so the added risk translates from something like 2.1 post-neonatal deaths per 1,000 live births in breast-fed babies to 2.7 per 1,000 in purely bottle-fed babies. It's not a huge worry for individual babies, or their parents, but over the 4 million or so births each year in the United States, that's up to 720 babies a year that needn't die.

Bottle-fed babies are also nearly twice as likely to develop eczema and asthma. They are at greater risk of developing childhood leukaemia – a cancer of the immune system. They are more likely to suffer from type 1 diabetes. They are also probably more likely to get appendicitis, tonsillitis, multiple sclerosis and rheumatoid arthritis. For parents, many of these risks are small enough not to need to worry too much. But again, as with the impact of bottle-feeding on infant mortality, the impact across the millions of babies born each year is significant enough that we *should* care.

Perhaps most significantly, feeding babies formula makes them more likely – perhaps as much as twice as likely – to become overweight. When scientists want to know if an effect is a true causation, not just a coincidental correlation, they look for 'dose dependency'. If one factor – let's say volume of alcohol consumed – is actually causing an effect – let's say slowed reaction time – we would expect the reaction time to be slower in line with higher volumes (extra doses) of alcohol, up to a point at least. The bigger the dose, the slower the reaction time.

The same relationship can be seen between breast-feeding and the risk of obesity. One study found that for each extra month of breast-feeding up to the age of nine months, the average risk of children becoming overweight drops by around 4 per cent. Two months of exclusive breast-feeding and the risk drops by 8 per cent; three months and it's about 12 per cent and so on. After nine months of breast-feeding a child is 30 per cent *less* likely to become overweight than a child who is bottle-fed from birth. Breast-feeding without any supplementation of formula appears to have an even greater effect, dropping a child's risk of becoming overweight by 6 per cent for each extra month. The impact of bottle-feeding on the likelihood of becoming overweight or obese is not limited to weight in childhood. Older children and adults remain more likely to be overweight as a consequence of how they were fed as babies. Obesity is often accompanied by type 2 diabetes, and bottle-fed babies are no exception. Being raised on formula alone makes a child 60 per cent more likely to go on to develop diabetes in adulthood. Just as with Caesarean sections, many of the risks of not breast-feeding relate to twenty-first-century illnesses.

For the baby boomers, born at a time when bottle-feeding was the norm, these facts and figures may be all too tangible. In the mid-Seventies, breast-feeding became fashionable once again, particularly among the well-off and well-educated. It may have been the inadvertent impact of formula companies' aggressive marketing campaigns in developing countries that led to the resurgence of breast-feeding in the late 1970s. Babies fed formula milk in some countries are up to twenty-five times more likely to die, largely because bottles cannot easily be sterilised and water supplies are often contaminated with pathogens. As women across North America and Europe railed against formula companies, breast-feeding rates soared. Within a decade, nearly three times as many women were breast-feeding their newborns.

But Generation X and the Millennials are not exempt from the risks brought by the bottle. Although in the past two decades breast-feeding in developed countries has continued to rise, up from around 65 per cent in 1995 to as high as 80 per cent in the past few years, breast-feeding behaviours still fall far short of official recommendations. For the 20–25 per cent of babies who never get a single gulp of breast-milk, and the next 25 per cent who are switched to formula in the first eight weeks, the rise in breast-feeding would not be much comfort. Even among those babies who are breast-fed beyond birth, half begin receiving supplementary formula within the first week.

In the United States, just 13 per cent of mothers manage the World Health Organisation's recommendations of six months of exclusive breast-feeding followed by breast-feeding supplemented by appropriate complementary foods for up to two years or beyond. In Britain, fewer than 1 per cent of mothers are still exclusively breast-feeding their baby by the time it reaches six months old.

Of course, breast-feeding is hard, particularly in the first few days and weeks. For some mothers there is no choice, and breast-feeding is just not an option, perhaps because of an unwell baby or a real problem with their milk supply. For other mums, financial pressures and a lack of support dictate the choices they must make. But as a society, we have perhaps lost sight of what's 'normal' when it comes to feeding our babies. In pre-industrialised traditional societies, babies are breast-fed for far longer than in the West. Weaning around two, three or four years old is typical, often until the next baby comes along.

The skewed Western perspective that comes along with 'Breast is best' (implying that formula is 'good') extends even to the scientific approach. Many studies ask 'What are the *benefits* of breast-feeding?' not 'What are the *risks* of bottle-feeding?' Statistically, these amount to the same thing: 'How do breast-feeding and bottle-feeding compare?' But, as Alison Stuebe, Assistant Professor of maternal–fetal medicine at the University of North Carolina School of Medicine, points out, the first question implies that breast-feeding is a bonus for a baby, perhaps like taking a multivitamin on top of an otherwise healthy diet. The second question implies that bottle-feeding is risky business – a step away from the norm. Breast-feeding is not the 'Gold Standard'; it is Standard. For women who are in two minds about how they feed their babies, this distinction could make all the difference to their choice.

This subtle difference in expression has a real impact on people's interpretation of the breast versus bottle debate. A survey in the United States in 2003 found that three-quarters of people disagreed with the statement: 'Infant formula is as good as breast-milk.' But only one-quarter of people agreed with the statement: 'Feeding a baby formula instead of breast-milk increases the chance the baby will get sick.' It seems there is a disconnect in what people know about the benefits of breast-milk, and what they think the consequences of not breast-feeding are. In a campaign targeted at women ambivalent about breast-feeding, those given advice about the 'benefits of breast-feeding' were less likely to choose to breast-feed than women given the same information but presented as the 'risk of not breast-feeding'.

Women should be free to choose how to bring up their own babies. But no mother should have to make these choices without access to honestly presented information about the impact of each option. As well as supporting women in their efforts to breast-feed, and making information more clearly and more appropriately available to women and health-care providers, babies would benefit from improvements to the quality of infant formula. At the moment, few formulas contain oligosaccharides or live bacteria. The trouble is, getting the perfect blend of the 130 different types, and including a healthy community of the most beneficial strains, is currently beyond us. Attempting it before we know the consequences could do more harm than good.

For the first three years of a child's life, its gut microbiota is highly unstable. Populations of bacteria come and go as they fight it out for territory. New strains invade and others retreat. A slow, steady decline in the abundance of the genus *Bifidobacterium* takes place over the first year or so of life. The greatest changes take place between nine and eighteen months of age, probably in sync with the introduction of new varieties of solid foods. In one experiment, introducing a baby to peas and other vegetables preceded a switch from a microbiota dominated by the phyla Actinobacteria and Proteobacteria to one composed of Firmicutes and Bacteroidetes. These dramatic shifts mark milestones in a baby's development.

Between eighteen months and three years, the gut microbiota looks more and more like that of an adult, gaining stability and diversity as the months go by. By a child's third birthday, the early differences in the microbiota that were brought about by either breast- or bottle-feeding have been swamped by the new strains picked up from other people and places. The lactic acid bacteria that were once so abundant become rarer, as the microbiota adapts to new foods and circumstances.

As a child ages, the microbes in its gut look less like a vaginal microbiota and more and more similar to its mother's gut microbiota. Like mother, like daughter – or son: if she has an oak forest, so will her children. In part this is because they live in the same house, surrounded by the same microbes, and they eat the same food. But it's also down to their shared genes. Remarkably, your genome has some control over which species it hosts. Genes involved in programming the immune system also influence the bacterial species that are allowed to live within the body. As mother and child share around half their genes, a child may benefit from having a matching set of microbes. After all, in the first few minutes of life, a baby's immune system has to cope with a massive invasion of bacteria, the like of which it will never encounter again. The fact that it can even survive what might have been regarded as a huge infection gives away that it has been genetically and immunologically forewarned. Having some preprogrammed intelligence about who is friend and who is foe probably goes a long way in helping the baby to cope with the onslaught of its mother's vaginal bacteria as it slips head-first into a microbial world.

The beauty of the microbiome is that, in ways that the human genome can never be, it is adaptable. As you age, as your hormones wax and wane, as you try new foods, as you visit new places, your microbes make the best of your situation. Poor nutrition? No problem – your microbes will help you to synthesise missing vitamins. Eating barbecued meat? Not to worry – your microbes will detoxify the charred bits. Changing hormones? That's fine – your microbes will adapt.

In adulthood, the body needs different amounts of vitamins and minerals than it did as a youngster. Babies, for example, need a lot of folic acid, but they can't eat the foods that contain it. Their microbiomes, though, are full of genes that synthesise folic acid from breast-milk. Adults don't need so much folic acid and they usually get enough from their diets, so instead of genes to synthesise this vitamin, their microbes contain genes that break it down.

The opposite is true for Vitamin B12. The older you are, the more you need. As you age, your microbiome increases the number of genes that synthesise B12 from food. Your microbes aren't doing this to be kind – they need these vitamins or their precursors too. Many other genes involved in synthesising or breaking down food molecules shift with age, making the most of the diet and coping with changes to the body.

Who you live with can also have a big impact on the microbes you carry. Just as you leave traces of your presence when you spend time in a house – fingerprints, footprints, DNA shed in skin cells and hair – you also leave a microbial signature behind. A study of the people and microbes residing in seven American households found that it was easy to identify which family belonged in which house, just by comparing the microbes on the human inhabitants' hands, feet and noses with the microbes on the floors, surfaces and doorknobs. Not surprisingly, the microbiotas of the kitchen and bedroom floors matched those of the family's feet, and the communities on kitchen surfaces and doorknobs matched those of their hands.

During the study, three of the families moved house. Within days, their new homes had been colonised by their bacteria, replacing those of the previous occupants. Indeed, a family member's contribution to the home's microbiota was so dynamic that if they spent even a couple of days away, their microbial trail went cold. It's possible that the steady decline in a person's microbial shadow could be used to create a timeline of events reliable enough for use in forensic investigations. DNA technology utterly changed crime-scene investigation, but your microbiome is even more distinctive than your human genome – imagine what secrets it could reveal.

Members of the same family tend to have very similar microbiotas to one another, and parents often share the exact same strains of bacteria with their children. Choosing to live with friends, or even strangers, though, could result in you sharing more than just the milk. One household included in the study was not a family, but three genetically unrelated people. Their shared environment was enough to bring about a merging of microbiotas. All three had many microbes in common, especially on their hands. Two of the housemates were in a relationship, and they shared more microbes with one another than either did with the third housemate.

For women, the waxing and waning of hormones each month can have a huge impact on the microbial make-up of the body. In many, the menstrual cycle is linked to abrupt shifts in the microbial species living in the vagina, with populations expanding and contracting in perfect synchronicity with each period. For others though, changes in vaginal microbiota composition appear to be completely random, showing no connection to the time of the month. Others still even have near-constant communities that seem utterly unaffected by menstruation or ovulation. Interestingly, among those women whose communities ebb and flow, the activity of the strains and species present often remains the same. A dominant lactic-acid-producing strain of *Lactobacillus* might suddenly disappear, but another lactic-acid producer, perhaps a friendly strain of *Streptococcus*, might replace it. So although the species have changed, the same jobs still get done.

I mentioned earlier that a woman's vaginal microbiota shifted in composition during pregnancy. The same goes for her gut microbiota. During pregnancy women gain between 25 and 35 lb in weight. Roughly speaking, that includes a 7 lb baby, plus a further 8 lb of placenta, amniotic fluid and extra blood. The remaining 10 to 20 lb is fat. In the final trimester, looking at the metabolic markers of a pregnant woman is a lot like looking at those of someone suffering from the health consequences of obesity. Excess body fat, high cholesterol, increased blood glucose levels, insulin resistance and markers of inflammation can all accompany both obesity and pregnancy.

But while in obesity these are all indicators of poor health, in pregnancy they mean something different. Changes in a woman's metabolism – her ability to process and store energy – are crucial to pregnancy. The extra layer of adipose tissue she gains, despite not actually needing to 'eat for two', probably provides the growing foetus with a safety net, ensuring that its mother has enough energy to support its growth. It also means she has the energy she needs to produce breast-milk once the baby is born.

Having learnt about the different microbiotas of lean and obese people, Ruth Ley of Cornell University wanted to know whether the microbial shifts behind obesity were also responsible for the metabolic changes in pregnancy. She and her team studied the gut microbes of ninety-one women as they progressed through their pregnancies. By their third trimesters, the women's microbiotas were profoundly different from the early days of pregnancy. They became far less diverse, and two groups – Proteobacteria and Actinobacteria – became far more abundant; changes that are reminiscent of those seen alongside inflammation in rodents and humans.

As I said in chapter 2, transplanting the microbiota of obese people into germ-free mice makes them rapidly gain body fat compared with transplanting the microbiota of lean people. These transplantations are a neat way to show that the microbes are *responsible* for weight gain, and not just a consequence of it. Ruth Ley took the same approach with the pregnant women's third-trimester microbiotas. Were they the cause or the consequence of the obesity-like metabolic changes of pregnancy? Germ-free mice given a transplant of human third-trimester microbes gained more weight, had higher levels of glucose in their blood and became more inflamed than mice given a microbiota from the first trimester. These changes could help collect and divert resources to the growing baby.

Once the baby is born, a woman's gut microbiota takes some time to return to normal, but return it does. For now it's not known exactly how long the pregnancy microbes hang around for, or what makes them shift back again, but it's intriguing to think that breast-feeding (and perhaps the hormones of labour) might have something to do with it. The effects of breast-feeding on 'shifting the baby weight' are well known, apparently using up the calories stored during pregnancy.

'Bubble boy' David Vetter suffered from the inherited disease Severe Combined Immunodeficiency and lived in an isolation bubble from his birth in 1971 to his death in 1984. He is the closest a human has come to living germ-free.

The release of the antibiotic penicillin to the public at the end of the Second World War meant an end to previously incurable infectious diseases, such as gonorrhoea.

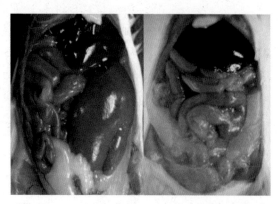

The caecum – an area of the gut usually densely populated with microbes – is grossly over-sized in germ-free mice (left) compared to normal mice (right). The reason for this dramatic anatomical difference is unclear.

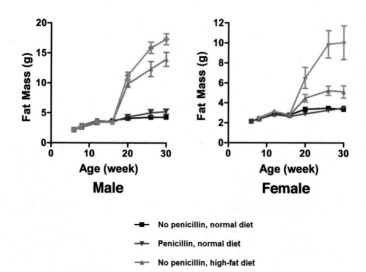

Mice fed a high-fat diet have a greater fat mass than those on a normal diet, but adding a daily low dose of the antibiotic penicillin from the start of life amplifies the effect of the high-fat diet, making the mice gain even more fat tissue.

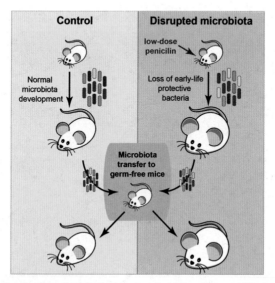

Transferring the gut microbes of mice treated with low dose penicillin from early in life into lean, germ-free mice results in weight gain in the recipient mice. The transfer of microbes from an untreated control mouse does not.

Anne Miller was the first person to be saved by penicillin, after she became critically ill with an infection following a miscarriage. Here, she stands with Sir Alexander Fleming (right) and an unidentified man in 1942, after her recovery.

Modern Western supermarkets are filled with boxes of processed and preserved food that is barely identifiable as either animal or plant, and is often far lower in fibre than its source.

Koala joeys do not have the genes necessary to digest eucalyptus leaves. By eating faecal 'pap' produced by their mothers, they build a colony of microbes that break down fibrous plant material for them.

After hatching, Kudzu bugs consume packages of microbes left by their mother next to the egg cases. If the packages are missing, the bugs wander in search of them.

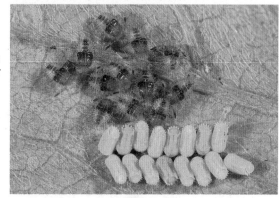

After a road accident, Peggy Kan Hai had surgery on her foot and developed a life-threatening antibiotic-resistant *Clostridium difficile* infection in her gut. She recovered after a faecal microbiota transplant from her husband.

Gastroenterologist Professor Tom Borody at the Centre for Digestive Diseases in Australia advertises to recruit stool donors for a clinical trial of faecal microbiota transplants in patients with inflammatory bowel disease.

Volunteers donating to Massachusetts-based stool bank OpenBiome save lives and earn money for each donated stool.

size of poop	# of people treated
50g	☠
100g	☠ ☠
150g	☠ ☠ ☠
200g	☠ ☠ ☠ ☠
250g	☠ ☠ ☠ ☠ ☠
300g	☠ ☠ ☠ ☠ ☠ ☠
350g	☠ ☠ ☠ ☠ ☠ ☠ ☠
400g	☠ ☠ ☠ ☠ ☠ ☠ ☠ ☠
450g	☠ ☠ ☠ ☠ ☠ ☠ ☠ ☠ ☠

THE MOST IMPORTANT THING YOU'LL DO ALL DAY!

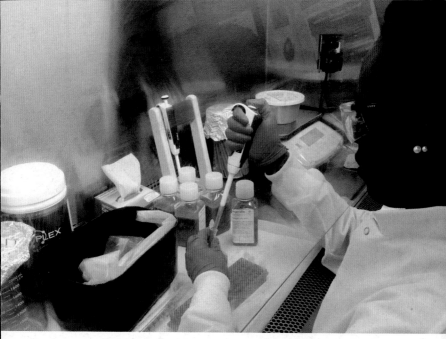

OpenBiome staff member Mary Njenga processing donated faecal samples to send to *C. diff* patients in hospitals and clinics across the United States for faecal microbiota transplantation.

Volunteers wishing to donate stool to OpenBiome are screened for microbiota-related health problems, including obesity, allergies, autoimmune diseases and mental health problems. Only a small proportion of volunteers are eligible to have their stool processed as a faecal microbiota preparation.

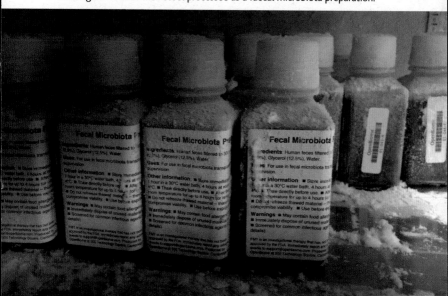

The bobtail squid relies on a single species of bacterium to provide a light source in its belly that camouflages it against the sky above, enhancing its chances of survival. Our own more complex microbial communities also benefit us, enhancing our health and well-being.

Whether breast-feeding also reverses the obesity-like microbial changes of pregnancy is not yet known, but we know that it does reduce women's risk of type 2 diabetes, high cholesterol, high blood pressure and heart attacks in later life.

Although it is affected by diet, hormones, travel abroad and antibiotics, the gut microbiota stays fairly stable throughout adulthood. Old age, though, alongside changes in health, brings changes to the body's microbial communities. As the body's human cells begin to show their collective age, so too do its microbial passengers. Individually, of course, very few human cells last the full duration of a lifetime, and most microbial ones stick it out for just a few hours or days. But as a superorganism, the human colony begins to run less efficiently and stumble more frequently as it ages. The immune system is largely responsible, having hoarded antibodies for decades. With age, people's immune systems become increasingly het up. The constant whirr of pro-inflammatory chemical messengers running through the bodies of the elderly is reminiscent of the low-level chronic inflammation seen in twenty-first-century illnesses. Termed 'inflamm-ageing', this medical characteristic of later life is closely tied to health.

Not surprisingly, it is also tied to the composition of the gut microbiota. Elderly people with a greater degree of inflammation and poorer health have less diverse gut communities, with fewer species that are known to soothe the immune system, and more that rile it. Whether age brings inflammation, and inflammation brings changes to the microbiota, or changes to the microbiota with age start the inflammation is not yet clear. But because diet in old age plays a big role in shaping the microbial community, it's likely that the microbiota are an important force in the ageing process. It's early days, but some scientists are even getting excited about the possibility of keeping people healthier for longer, or even extending the human lifespan, by altering the gut microbiotas of the elderly.

We are accompanied from the very first breath to the very last by our colony of microbes. As our bodies grow and change, our microbiomes adapt, providing us with an extension to our own genomes that can adjust within hours to better suit our needs, and their own. All being well, a mother's microbes make the best birthday present a child could wish for. Our parents' choices live with us, quite literally, as we learn to walk, to talk, and to care for ourselves. As adults, the responsibility of caring for all the cells in our bodies, both human and microbial, falls to us. As mothers, women pass on not only their own genes, but the genes of hundreds of bacteria. The genetic lottery of life has an element of chance, but also one of choice. The more insight we gather into the importance and the consequences of a natural birth, and extended, exclusive breast-feeding, the more empowered we will be to give both ourselves and our children the best chance of lives of health and happiness.

EIGHT

Microbial Restoration

On the evening of 29 November 2006, as 35-year-old counsellor Peggy Kan Hai drove through the rain to meet a client on the island of Maui, in Hawaii, she was hit by a motorcyclist travelling at 162 mph. Pinned inside the wreckage of her car, bleeding from her head and mouth, she slipped in and out of consciousness. The young man who had hit her died on the road amongst the debris of his bike.

In 2011, after five years of surgeries to repair injuries to her head and legs, Peggy's damaged left foot became necrotic. With her life at risk from the spreading sepsis, Peggy had no choice but to have surgery to partially amputate her foot, and fuse the bones of her ankle together. Three days after the operation, Peggy fell violently ill with sickness and diarrhoea. Her nurse was horrified and summoned the surgeon the following day. He assured Peggy it was just a reaction to the medication – anaesthetics, antibiotics and pain-relievers – that she'd been given. He sent her home that evening with different drugs.

A few weeks later, Peggy gave up on the drugs despite the pain in her foot, in the hope that things would improve. They did not. The following morning marked the start of two months of up to thirty episodes of diarrhoea a day. Peggy lost 20 per cent of her body weight, her hair fell out, her nerves buzzed and her vision blurred. Her doctor insisted she had normal symptoms of opiate withdrawal, then decided it was irritable bowel syndrome or acid reflux. Peggy refused diarrhoea medication, feeling certain that bottling up the cause of her illness would only make her worse.

Several months later, Peggy was sent to see a gastroenterologist at the hospital where she'd had her foot surgery. After a colonoscopy (viewing the inside of the colon with a tiny camera), an explanation for Peggy's severe diarrhoea emerged. She had *Clostridium difficile*, or *C. diff*.

C. diff is the particularly nasty bacterium I mentioned in chapter 4, which can cause an awful, life-threatening infection. It lurks steadfastly in hospitals and the healthy guts of humans. It has a couple of clever tricks that give it the edge over other microbes and hospital cleaning staff alike. For starters, in the last few decades a new strain has evolved which is both more resistant and more dangerous than previous versions. It's quite likely this mutated strain has emerged in response to the ongoing arms race between the bacterium and the antibiotics we throw at it and its cousins. At the moment, *C. diff* is in the lead.

Its other trick is one that it holds in common with up to a third of the bacteria that live in the gut, and many pathogens besides: it can form spores. Like a frightened armadillo curled tightly into an armoured shell, *C. diff* packages itself into a thick protective layer to survive when times get tough. Antibacterial cleaners, stomach acids, antibiotics and extremes of temperature all wash over these spores, allowing them to persist until the danger has passed.

Peggy's circumstances were typical. She had been given antibiotics during her foot surgery, and had then spent several days in a hospital – a stronghold for *C. diff*. The antibiotics, as well as protecting her from an infection in her wounds, had disrupted her gut microbiota, leaving her vulnerable to invasion by *C. diff*. Like a weed, it had rampaged through the landscape of her intestine before the protective layer of beneficial microbes could re-establish their colonies. Peggy's gastroenterologist prescribed course after course of high-dose antibiotics to rid her of the *C. diff* infection, but it held fast, only making her sicker.

As Peggy's vision and hearing deteriorated, and her weight dropped dangerously low, she and her husband realised they needed to take drastic action to restore the microbes of her gut, and oust the *C. diff*. The question was, how?

Peggy Kan Hai's dilemma is not unique to sufferers of *C. diff*. For many others suffering diseases of the digestive system, and other conditions stemming from a damaged microbial ecosystem, the question of how to undo the harm and reinstate a healthy colony of microbial friends is a pertinent one. No doubt, a good diet and the avoidance of unnecessary antibiotics are fundamental to maintaining a healthy microbiota, but what if your colony has already been decimated? What if key species have long since departed, and opportunists have taken their place? What if the immune system has lost sight of who are the enemies and who are the allies? Tending the ruins of a once-thriving microbial community may do little more good than it would to sprinkle water on the parched brown twigs of an abandoned house plant. Sometimes the only option is to start again: prepare the ground and sow new seeds.

In 1908, Elie Metchnikoff published a book with a title showing positivity uncharacteristic of this Russian biologist. This was a man who had twice attempted suicide – once through an overdose of opium, and the second time by deliberately infecting himself with relapsing fever, in an effort to become a scientific martyr. Yet his third book, *The Prolongation of Life: Optimistic Studies*, was concerned not with hastening death, but with delaying it. Perhaps it served as an attempt to lift his ageing spirits as he approached a final, and this time inevitable, confrontation with his own mortality. Metchnikoff, who won a Nobel Prize for his work on the immune system in the same year as the book was published, shared with his ancient predecessor, Hippocrates, the notion that death sits in the bowels. From his relatively modern and enlightened perspective, Metchnikoff suspected that it was the newly discovered microbes of the intestines that were the true seat of senility.

From the methodological hilltop of twenty-first-century science, reading Metchnikoff's treatise is an alarming, and at times amusing, experience. Though his hypothesis is an interesting one, it runs low on evidence, and includes such pseudo-correlative nuggets as the idea that bats have no large intestine, very few microbes, and yet far outlive other small mammals. It is, he surmises, the presence of microbes, and the large intestine housing them, that brings early death in more densely populated mammals. So why is the large intestine there? he pondered. 'In answer to the question, I have formed the theory that the large intestine has been increased in mammals to make it possible for these animals to run long distances without having to stand still for defecation. The organ, then, would simply have the function of a reservoir of waste matter.'

Metchnikoff was not alone in his ideas about the gut's microbes causing ill-health. A major new hypothesis regarding the cause of a multitude of diseases, both physical and mental, was doing the rounds among doctors and scientists. Named 'autointoxication', its central tenet was that the colon was, in the words of one French doctor, 'a receptacle and laboratory of poisons'. The bacteria of the gut were thought to simply be rotting the remains of food, and creating toxins that brought about not only diarrhoea and constipation, but fatigue, depression and neurotic behaviour. In cases of mania or severe melancholy, surgical removal of the colon was often prescribed – known as a 'short-circuit' procedure. Despite a frighteningly high death rate, and a huge impact on quality of life, this radical intervention was deemed worthwhile by the doctors of the day.

Far be it from me to critique the degree of adherence to the scientific method of a Nobel prize-winner, but Metchnikoff's dabblings in intestinal microbiology, in this book at least, barely met reasonable standards of repeatability, comparison against a control, or concerns of causation. His scientific coming-of-age coincided with a period of history in which medical scientists were overcome with excitement about the research avenues opened up by Louis Pasteur's germ theory. Hypotheses thrived, and little time or mental energy was devoted to patient study, experimentation or evidence-building before the new cohort of medical microbiologists bounded off, tails wagging, to sniff out new ideas.

Nonetheless, the media, the public and a slew of charlatans jumped on the autointoxication bandwagon in the early twentieth century. Aside from inflicting colon surgery, a couple of other treatments for bad bacteria emerged. One was colonic irrigation, still favoured by so-called medi-spas today, and not well-regarded by the medical community. The other was to eat a daily dose of good bacteria; what we now call probiotics.

Metchnikoff's musings on prolonging human lifespan were prompted by a rumour he heard from a Bulgarian student. It was said that among the peasants of Bulgaria were a large number of centenarians, and that their secret to a long life was a daily drink of soured milk – yogurt. Of course, the sour taste of the fermented milk came from the lactic acid produced when bacteria that Metchnikoff referred to as 'Bulgarian bacillus' fermented the milk's lactose sugar. Now, this strain of bacteria is classified as *Lactobacillus delbrueckii* subspecies *bulgaricus*, though it's often referred to as *Lactobacillus bulgaricus*. Metchnikoff believed that these lactic acid bacteria were disinfecting the gut, killing the harmful microbes that ushered in senility and death.

Tablets and drinks containing *Lactobacillus bulgaricus*, and another strain, *Lactobacillus acidophilus*, were soon available in the shops. Adverts declaring amazing results filled medical journals and newspapers. 'The results are nothing short of amazing. Not only a banishing of mental and physical depression, but a flooding of new vitality throughout the system,' claimed one brand. Autointoxication was soon widely accepted by doctors and the public, and in the first few decades of the twentieth century, the probiotics industry took off.

But it didn't last. The theory of autointoxication was a scientific house of cards, upon which the increasingly top-heavy probiotics industry rested. The layers of interconnected hypotheses that formed its structure each had merit, but only the merest hints of evidence to cement them in place. Just as with the promising new research into the role of pathogenic microbes in causing serious mental health conditions, it would once again be Freud and his followers who blew down this particular structure. Ironically, it would be replaced by a far more harmful house of cards in the form of psychoanalysis and the Oedipus complex.

A key player in toppling the theory of autointoxication was a Californian doctor named Walter Alvarez. Using little more evidence than had backed the microbial ideas of Metchnikoff, Alvarez embraced all things psychoanalytical. He branded all patients who raised the idea of autointoxication as psychopathic, dismissing them after an initial consultation. Rather than taking a medical perspective, he tended to draw his diagnoses from the character and appearance of his patients. Women suffering migraines, for example, were deemed by Alvarez to be found amongst those who had small, trim feminine bodies with well-formed breasts. He advised his fellow doctors to look out for such women and check their symptoms accordingly. Even the most basic of gastrointestinal complaints – constipation – was no longer seen as a consequence of troublesome microbes by the physicians of the day, but as chronic hypochondria combined with ano-erotic fixations.

The science of autointoxication undoubtedly lacked rigour at the time, in no small part because the microbiologists of the day lacked tools. Nevertheless, among the charlatans, the colon-cleansers and the yogurts of dubious microbial origin were some good ideas. It took until 2003 for the value of probiotics in mental health conditions to once again be discussed by a brave group of scientists. But by then they had DNA-sequencing technology, a system of scientific peer-review, and a research climate free from the guilt-laced clouds of Freudian thinking.

Probiotics, then, despite managing to sustain a presence on the supermarket shelves in both their food and tablet forms, did not re-enter the scientific consciousness until recently. Once again, the probiotics industry is a burgeoning one, and a handful of brands are household names. Without actually promising it, manufacturers of many a little yogurt pot use clever marketing to suggest you will feel brighter, smarter, fresher, less bloated, more awake, happier and healthier if you down a *Lactobacillus*-enhanced drink or two each morning. Brands compete over the strains their products contain, and their supposed benefits. Patent applications claim the right to produce and market the particular combinations of genes and strains that give each variety of probiotic its own special powers. *Lactobacillus rhamnosus* plus *Propionibacterium*, for example, to crowd out *E. coli* 0157. Or *Lactobacillus* combined with 'dialkylisosorbide' to combat acne. How about a vaginal insert comprised of nine *Lactobacillus* species and two *Bifidobacterium* species to control imbalanced vaginal pH? And what about a very specific genetic variant of *Lactobacillus paracasei* for pregnant women to prevent allergy in their newborn baby?

It's all very well, but despite the patent applications, drug-regulation rules in most countries don't currently allow for products that contain bacteria to make health claims. What were once just fermented foods and dietary supplements are now beginning to seem suspiciously like medicines, thanks to the scientific research into the health benefits of the live bacteria they contain. Of course, if *Lactobacillus* species really can prevent *E. coli* infections, cure acne or prevent allergy, yogurt manufacturers are going to want their customers to know. But true medicines have to go through an expensive battery of clinical trials before they're unleashed on the public. In theory at least, drug companies must establish that their medicines are both effective and safe. Clearly eating yogurt is safe, but what about effective? Can probiotics actually make you healthier or happier?

Technically, the answer is a resounding yes, but that's because the very definition of probiotics, according to the World Health Organisation, is that they are 'live microorganisms which, when administered in adequate amounts, confer a health benefit on the host'. Asking 'Do probiotics really work?' is a tautologous question. The real question, then, is which bacteria, and how many of them, can prevent, treat or cure disease?

I'd like to blow you away with stories of miraculous recoveries brought about by little more than a pot of yogurt or a freeze-dried colony of friendly bacteria. I'd like to tell you that *Lactobacillus inventedus* will cure your child's hay fever, and that *Bifidobacterium fantasium* will help you lose weight. But of course it's not that simple.

In your gut are 100 trillion microbes. 100,000,000,000,000. That's about 1,500 times the number of humans living on the planet, all squeezed into your belly. Among those 100 trillion microbes are perhaps 2,000 different species. That's about ten times more species than there are human nationalities. And within those 2,000 species are countless different strains, all with an arsenal of different genetic capabilities. Yes, they are mostly 'friendly' from *your* perspective, but among one another it's not always so neighbourly. Populations vie for space, crowding out weaker opponents. Species defend their patch through chemical warfare, killing those that dare to invade. Individuals compete for nutrients, growing tails to propel themselves into more profitable territories.

Now imagine adding a little pot of yogurt to these badlands. Imagine the little band of tourists, swimming among the milk and the sugar of their yogurty vehicle. Perhaps 10 billion or so of them, looking for a place to settle. It might seem like a lot, but that's four zeros fewer than the home crowd. Hardly an impressive army to take into such a battleground. Like baby turtles venturing into the vast ocean for the first time, many are probably picked off in the first splashes of freedom from their tiny plastic bottle. For those that reach the gut – which they certainly can do – the challenge is to set up shop and make a decent living. It's not an easy one in an already crowded and not particularly welcoming neighbourhood.

Not only are they vastly outnumbered, but the combined skill-set of these brave tourists is a small one. They all belong to the same strain of bacterium, with the same genes, and therefore the same means. In comparison with the 2,000 or more species your gut contains, and the 2 million or more genes they carry, the tourists, probiotic or not, have a limited repertoire of tricks up their sleeves. Ultimately of course, what bearing their particular tricks have on the health of their host – us! – matters as much as any other obstacle on their journey to bringing the benefit implied in their status as probiotics.

Before I get sued, though, let me tell you what those tourists who do manage to hang around long enough and in large enough numbers to have an effect have to offer. I'm not just talking about yogurts here, but the more medicinally-styled pills, bars, powders and liquids that contain live bacteria, sometimes more than one species.

Let's start with the most basic expectation you might have of a *pro*biotic: that it would make up for the most unpleasant side effect of *anti*biotics. Decimation of the microbiota's ranks is often the unintended consequence of trying to eliminate a bug with anti-biotics. For many people – about 30 per cent of patients – the result of that microbial depletion is diarrhoea. It's called 'antibiotic-associated diarrhoea', and it usually goes away again once the course of drugs is finished, unless you're unlucky enough to contract something like *C. diff*, as Peggy Kan Hai did after her foot surgery. If it's sim-ply the loss of good bacteria that prompts the diarrhoea, immediately replacing them with more good bacteria should halt, or at least im-prove, the symptoms.

And so it does. There's no major leap of faith involved in believing that among sixty-three well-designed clinical trials, including a total of nigh on 12,000 participants, it was found that probiotics significantly reduce the chance of developing antibiotic-associated diarrhoea. Of the 30 in 100 people who would normally get diarrhoea, only 17 will get it if they are also taking a probiotic. Exactly which bacteria, and what dose is most effective, isn't as easy to tell. It's quite likely that particular antibiotics are more prone to inducing diarrhoea, and knowing which those are could make a tandem probiotic prescription just the ticket for avoiding this side effect. It's a worthy use of pro-biotics; considering that around 8 million Americans are taking antibiotics at any one time, over 2 million of them are probably suffering diarrhoea, and nearly 1 million needn't be, if only they had the right probiotics to prevent it.

Probiotics also pull their weight in the tiniest of babies. Sometimes when babies are born too early, their intestines begin to die. Giving these premature babies preventative probiotics reduces their risk of dying by 60 per cent. In babies and children with infectious diarrhoea, probiotics – particularly one strain known as *Lactobacillus rhamnosus* GG – reduced the duration of the illness.

But what about more complex illnesses? What about conditions that have taken root already? For fully-fledged autoimmune and mental health conditions, like type 1 diabetes, multiple sclerosis and autism, the trouble with probiotics is probably that they are too little, too late. The insulin-releasing cells of the pancreas have already shut up shop, the nerve cells have been stripped of their sheaths, and the developing brain cells have been pushed off course. Even in allergies, although no cells are destroyed, the immune system is already out of control. Bringing it back in line may be as difficult as awakening the cells of the pancreas or re-coating the nerves.

Real, well-designed, peer-reviewed studies have found that probiotics of various brands, species and strains can indeed make you happier and healthier, improving mood, alleviating eczema and hay fever, lessening IBS symptoms, preventing diabetes in pregnancy, curing allergies and even encouraging weight loss. Whilst most of these conditions are not cured altogether, at least not with just a few weeks or months of probiotic treatment, they do carry some benefit. To see a real effect of probiotics, though, prevention is certainly proving to be better than cure.

Take this experiment in mice, for example. There's a breed that, by a quirk of the genes, is almost guaranteed to develop the mouse equivalent of type 1 diabetes by the time it reaches adulthood. But give these mice VSL#3 – a probiotic composed of 450 billion bacteria of eight different strains – every day from four weeks old, and this genetic 'fate' is diminished. Whereas 81 per cent of the mice given a placebo had diabetes by the age of thirty-two weeks, only 21 per cent of those given VSL#3 did. Three-quarters were protected from near-inevitable autoimmunity by a single daily dose of live bacteria.

Starting the mice on VSL#3 a little later in their lives, at ten weeks, showed that it's a case of better late than never. By thirty-two weeks, around 75 per cent of placebo-treated mice were diabetic, but only 55 per cent of probiotic-treated mice had developed the disorder. Not as impressive as the results from four weeks of age, but still a significant reduction in diabetes cases.

The hundreds of billions of bacteria contained within VSL#3, which claims to have more individuals and more species than any other marketed probiotic product, are somehow altering the diabetic disease process in these genetically susceptible mice. Whereas normally the immune system would take a dislike to the insulin-producing cells of the pancreas, these bacteria seem to prevent this. The immune systems of VSL#3-treated mice seem to marshal a team of white blood cells that march over to the pancreas, where they pump out an anti-inflammatory chemical messenger that prevents the destruction of the pancreatic cells. It's inspiring stuff: perhaps a judiciously timed course of probiotics would halt these diseases before they ever got going in humans too? There is a clinical trial under way to test exactly that idea, but results will not emerge for some time yet.

At the core of it, probiotics must have some kind of effect on the workings of the immune system to be beneficial to health. Going back to the underlying origin of twenty-first-century illnesses, it's the *inflammation* that plagues our bodies that probiotics need to see off if they are to have true value. Remember the T regulatory cells (T_{regs}) in chapter 4? They were the brigadiers of the immune system, calming down the bloodthirsty soldier cells when there was nothing to attack. Ultimately, these brigadiers are controlled by the microbiota, which recruit more, and better, T_{regs} to prevent the immune system launching an attack on them. Probiotics mimic this effect, encouraging the existing T_{regs} to suppress the rebel members of the immune system's ranks. Once again, VSL#3 has a beneficial effect in mice, decreasing the leaky-gut effect that seems to be both cause and consequence of inflammation.

Three things matter when it comes to probiotics. First, what species and strains does a product contain? Often, these are not detailed, or they don't match up with the true contents when they're cultured or sequenced. Probably, the more species the better, though we have very little understanding of the impact of different strains on the body. Second, how many individual bacteria, or 'colony-forming units' (CFUs) does a product contain? This goes back to competition these tourists face on their way through the gut. The more CFUs, the more chance of having an effect. Third, how are the bacteria packaged? Probiotics come in all forms: powders, pills, bars, yogurts, drinks and even skin creams and washes. Some are mixed with other supplements – a multivitamin, say. What these preparations do to the bacteria is simply not known. Many probiotics in yogurt form are accompanied by a fair dose of sugar, which may even tip the balance towards making them more unhealthy than healthy.

The first of these points – the species and strains – is probably the most contentious. Metchnikoff's legacy lives on in many of the strains that are usually marketed as probiotics. Members of the genus *Lactobacillus*, for all their value in yogurt-making, are not that numerous within the human adult gut microbiota. They thrive in the guts of vaginally delivered, breast-fed babies, yes, but once their work is done, their collective abundance drops to less than 1 per cent of the total bacterial gut community. Lactobacilli got a head start in the probiotics industry for one main reason: they can be cultured. Because they can survive in oxygen, unlike most of the gut microbiota's members, they are relatively easy to grow on a Petri dish, or, indeed, in a tank of warm milk. This means that they were over-represented in the earliest studies of the human flora. Had probiotics taken off in the brave new world of DNA sequencing and anaerobic (oxygen-free) culturing, it's very unlikely we would have selected lactobacilli as the group best suited to fortifying our diverse colonies.

Probiotics have their place, but what if you're in a situation like

Peggy Kan Hai's? With her weight plummeting and no remaining antibiotic options, Peggy was desperate. The risk of developing 'toxic megacolon' weighed on her mind; her colon could effectively explode, releasing its contents into her torso. If that happened, her chances of death were horribly high. In the past year, around 30,000 people had died from a *C. diff* infection in America – far more than die of AIDS, for example – and Peggy did not want to join them.

There was one other treatment option available. Peggy had heard through a friend whose sister worked as a hospital nurse that some patients with untreatable diarrhoea were being given a new therapy, available only at a handful of hospitals worldwide. Apparently, they were getting better. Peggy was willing to try anything. A few phone calls to one such hospital later, and Peggy was booking flights from Hawaii to California for the treatment. Her husband would accompany her, but not just for moral support. It was he who would provide Peggy with the donation she badly needed – a new set of gut microbes. That those microbes were to be found in his faeces did not deter either of them; there were simply no other options left.

Known as Faecal Microbiota Transplantation, Bacteriotherapy, or, my favourite, a Transpoosion, this is exactly as it sounds. Take faeces from one individual, and put them in the gut of another. It sounds disgusting, but we are not the first species to have the idea. Other animals, from lizards to elephants, indulge in coprophagy now and then. For some, like rabbits and rodents, eating their own faeces is an essential part of the diet, as it allows them to have a second go at the nutrients locked away inside plant cells once their gut microbes have cracked them open. It's not a trivial contribution to their caloric intake, either. Rats only grow at three-quarters of the normal rate if they are prevented from eating their own poo.

For other species, though, coprophagy is relatively unusual, and is often labelled as 'abnormal behaviour' by zoologists. The matriarchs of herds of elephants, for example, have been seen to produce a runny dung, apparently so that younger members of the herd can scoop it up with their trunks and eat it. Chimpanzees, too, will eat one another's faeces. According to the pre-eminent British zoologist Dame Jane Goodall – whose dedicated studies of the chimpanzees of Gombe Stream National Park in Tanzania led to a revolution in our understanding of chimp behaviour – some wild chimps turn coprophagic when they have diarrhoea. A glut of newly ripened fruit in the forest can result in a bout of diarrhoea as chimps' gut microbiotas adjust to the new food source. One individual that Goodall studied, a female called Pallas, suffered from chronic diarrhoea on and off for ten years. Whenever the diarrhoea returned, Pallas would go coprophagic. From our new-found microbial perspective, it's tempting to speculate that Pallas was using the faeces of healthy chimps to restore her own microbial balance. Perhaps chimps suffering after indulging in a new fruit crop are coprophagic because it allows them to acquire microbes from other members of the troop who may have been through a few more fruiting cycles than they have, with the bugs to prove it.

Though it's rare in the wild, zoo animals are particularly keen on coprophagy, much to the delight of small children, and the chagrin of zoo keepers. The practice is often put down to boredom, along with stereotypical behaviours like rocking, pacing and obsessive grooming. A psychiatrist used to dealing with patients with autism, Tourette's syndrome and obsessive–compulsive disorder might notice some similarities in these behaviours between their patients and captive animals. Not least, the fascination with faeces – both eating them and smearing them – that the most severely autistic children, as well as some schizophrenic and OCD patients, display. The Freudian interpretation for both the animals' and the patients' repetitive and coprophagic behaviour might be one of parental estrangement or psychosexual frustration. The physiological interpretation, though, puts a bit more of a microbial spin on it: what better way to correct an aberrant microbiota that's producing repetitive behaviour than to consume the faeces of another, more healthy, individual? Coprophagy, then, would not be an abnormal behaviour, but an adaptive one – a sick animal trying to correct its dysbiosis.

Indeed, in experiments, providing captive chimps with fibrous leaves reduces their coprophagic behaviour. They don't actually eat the leaves; rather, they suck on them, and tuck them under their tongues. It's just a guess, but perhaps they are sucking off the layer of bacteria that makes its living digesting those leaves? That way they can seed their own microbiota with bacteria, or bacterial genes, that help them to digest their food, much like the Japanese people I mentioned in chapter 6 whose microbiotas contain genes from bacteria living on seaweeds used in sushi. With the beneficial microbes from the leaves on board, perhaps coprophagy is less important for the captive chimps – eating their meals just the once is now enough.

Providing germ-free laboratory mice with a new microbiota is easy – just house them with mice that have one already. A few days of coprophagy later, and the mice have matching microbes. So much so that even housing two groups of mice that both have a full set of microbes can result in changes in the microbial species they harbour. In yet another clever experiment led by Jeffrey Gordon at Washington University in St Louis in 2013, researchers took two sets of germ-free mice, and inoculated one set with the gut microbiotas from obese humans. The twist in this experiment was that each of these obese humans had a twin, and their twins were all lean. So the second set of germ-free mice received the gut microbiotas of the lean twins. As expected, the mice with the obese microbiotas gained more body fat than those with lean microbiotas. Five days after their inoculations, the two sets of mice were brought together: a mouse with the microbiota from an obese twin was housed with a mouse with the microbiota from the corresponding lean twin. Remarkably, the obese mice gained less weight living alongside their lean 'twins' than they did if they were not exposed to these lean microbes. Checking the microbiotas of the two mice revealed that the obese mouse's microbiota had shifted towards that of the lean mouse, whereas the lean mouse's had remained stable.

If we were chimpanzees, we might indulge in a little faecal sharing to keep ourselves lean and healthy. Fortunately though, to benefit from a fellow human's healthy microbiota, coprophagy is entirely unnecessary. That's not to say that our more clinical alternative – faecal transplant – is entirely palatable though. In its crudest form, it involves mixing stool from a healthy donor with some saline, whizzing it up in a kitchen blender, and squirting it out into the large intestine of the patient through a long plastic tube with a camera attached – a colonoscope – inserted from the bottom up, so to speak. Occasionally, faecal transplants are delivered from the top down via a nasogastric tube that runs into the nostril, down the throat and into the stomach.

One of the pioneers of the modern use of faecal transplants, Dr Alexander Khoruts, recalls his early experiences of preparing faecal suspensions: 'I did the first ten transplants the old-fashioned way, with a blender in the endoscopy bathroom. During that experience I quickly appreciated the practical barriers to doing faecal transplantation in a busy clinical setting. The olfactory potency of human faecal material revealed at the touch of a button on the blender can be quite shocking – it can empty the waiting rooms.' Not only that, but making an aerosol out of faeces, however pathogen-free they are, is probably not particularly safe for the doctor preparing the suspension. Even the most well-meaning of bugs can be harmful if it gets in the wrong place – what's healthy in the gut may not be in the lungs.

The idea of it is disgusting, isn't it? If you're still reading, let me get this out of the way. There are two ways for me to deal with the yuck-factor of faecal transplants. One is to gloss over it, be euphemistic about it, and hope you don't think about it too hard. The other is to confront its disgustingness. Yes, it looks disgusting. But it is just microbes, dead plants, and water. Mostly it is bacteria – around 70 per cent of it or more. The brown colour comes from the pigment from broken-down red blood cells, which is converted by your liver and ejected as waste. And yes, it smells disgusting. But those smells are just gases, mainly hydrogen sulphide and other sulphur-containing gases, created by your gut microbes as they break down the remains of your food.

Disgust is a protective emotion. It has evolved because it keeps us away from things that harm us. Vomit, rotting matter, swarms of insects, the bodies of people we don't know or love, slimy things, sticky things, gunky things. And faeces. We are particularly disgusted by the faeces of meat-eaters – which would you rather touch, a dog poo or a cow pat? – and those of fellow humans. The world over, we all make the same face when confronted with something revolting; our heads pull back, our noses flatten, our brows furrow. We draw our hands into our chests and turn away. If it's really repulsive, we vomit instantly. This evolutionarily-encoded disgust reaction helps us to avoid coming into contact with pathogens that might make us sick. They could be in the vomit, the rotting matter, the slimy thing, the sticky thing. They could be in the faeces.

So not wanting to think about poo, and least of all someone else's poo in your own body, is perfectly natural. Instead, think for a moment about having a blood transfusion. Probably you're not quite as disgusted by that idea. Bags of blood, carefully collected from healthy donors, and screened for any diseases that might be stowed away in the cells or plasma. Labelled with their blood groups and collection date, hanging up, saving lives. It's quite a clinical, sterile, almost futuristic image.

But blood, like faeces, can carry pathogens, such as HIV and hepatitis. Blood, like faeces, rots when it's left exposed to bacteria in the air. And *faeces*, like blood, can be life-saving. Think of a faecal microbiota suspension as it really is: a liquid with health-giving properties. Dr Alexander Khoruts tells the story of a medical student who came to donate her stool for use in *C. diff* patients. When she told her friends, many of whom were also medical students, instead of praising her selflessness and muttering that they really ought to find time to donate as well, as they might had she been giving blood, they laughed and teased her for her efforts.

Twenty-first-century doctors, aware of the emerging science of the microbiota, were not the first to discover the life-saving properties of stool. A traditional Chinese medicine doctor from the fourth century by the name of Ge Hong wrote in his *Handbook of Emergency Medicine* that giving patients with food poisoning or severe diarrhoea a drink made from a healthy person's stool would bring about a miraculous cure. The same treatment is mentioned 1,200 years later, again in a Chinese medical handbook, and is this time referred to as 'yellow soup'. Evidently, making faecal transplant palatable to patients, metaphorically speaking, was as difficult then as it is now.

Except that it's not that hard to persuade someone who's spent the past three months on the toilet, and who's lost a fifth of their body weight, to give faecal transplant a go. For Peggy Kan Hai, any 'yuck' factor that she might have felt about a faecal transplant before she fell ill had disappeared. In the clinic in California, recovering in the hours after her colonoscopy and the delivery of the suspension of her husband's filtered faecal microbes, Peggy was already better. For the first time in months, she did not need to go to the toilet. For a full forty hours she remained out of the bathroom. A few days later, the diarrhoea was gone. After two weeks, her hair began to grow again, the acne on her forty-year-old face began to clear up, and the lost weight began to return.

Treating recurrent *C. diff* infections with antibiotics has about a 30 per cent cure rate. Over a million people are infected each year, and tens of thousands die. But treating *C. diff* with a single faecal transplant has a greater than 80 per cent cure rate. For those who relapse after the first transplant, as Peggy later did, a second transplant brings the cure rate up to over 95 per cent. It's hard to think of any other life-threatening disease that can be treated in a single non-surgical procedure, without the need for drugs, at a cost of just a few hundred dollars, with such a high success rate.

For the gastroenterologist Professor Tom Borody, faecal transplants have become a mainstay of the treatments he offers at the Centre for Digestive Diseases in Sydney, Australia. In 1988, Borody had a patient, Josie, who had picked up a bug on holiday in Fiji. She had suffered ever since from diarrhoea, cramping, constipation and bloating. What should have been an easy fix with a course of antibiotics dragged on and on until Josie became suicidal. Borody was distressed by his patient's plight, and was running out of options to restore her to her pre-Fiji state of health. He delved into the literature, and discovered the cases of three men and one woman who, in 1958, had developed severe diarrhoea and abdominal pain after being treated with antibiotics – a similar situation to Josie. Three were at death's door in intensive care, and the statistics didn't bode well – a 75 per cent mortality rate was standard for their condition back then. Their doctor, one Ben Eiseman, had treated them using a faecal transplant. Within hours or days of receiving faecal enemas, all four patients were able to get up and walk out of hospital – completely free of the diarrhoea that had plagued them for months.

Borody, excited by the possibility of a cure for his patient, suggested the idea to her. She, like Peggy Kan Hai, was willing to try anything. Borody administered the transplant over two days. A few days later, Josie was dramatically better – so much so, that she was already able to go back to work. Borody, meanwhile, didn't dare tell anyone what he'd done to help her, such was the resistance to the concept at the time. But from that point on, he and his team started using faecal transplants to treat any condition that they thought might benefit from a microbial restoration. Over the next year, they carried out fifty-five faecal transplants for complaints ranging from diarrhoea to constipation and inflammatory bowel disease. Twenty-six patients got no better, but nine improved, and twenty recovered completely.

Over the coming years, Borody and his team learnt which conditions would respond to faecal transplant and which would not. They have now carried out over 5,000 transplants, most of which were for diarrhoea-predominant irritable bowel syndrome and *C. diff* infections. With a cure rate of about 80 per cent in Borody's clinic, faecal transplant is by far the most effective therapy for this form of IBS. Constipation is harder to treat, with a cure rate of only about 30 per cent, and may take many days of repeated transplants. Despite the successes, and the patient demand, Borody and other doctors who use faecal transplants are often accused of being charlatans, even by other leading doctors. Because stool is not a drug that can be manufactured and sold, these transplants are not subject to the rules and regulations of other medicines. Clinical trials are not strictly necessary, and as such, many doctors are sceptical that faecal transplants are genuinely effective.

But Peggy Kan Hai, and others in her position, have been quite willing to embrace the idea. As Peter Whorwell, Professor of Medicine and Gastroenterology at Manchester University, puts it, 'My IBS patients are gagging for it.' In reality, it is often the doctors who are holding back, either because of the inherent yuck factor or because they view faecal transplant as a bogus therapy. In America, even the regulators at the Food and Drug Administration (FDA) have tried to put a stop to the use of faecal transplants in medical clinics. For a critical two months in spring 2013, the FDA banned the treatment for all but a handful of approved medics. Doctors who had been successfully treating patients for *C. diff* and other digestive diseases suddenly faced needing to apply for a new licence. The FDA were concerned about the safety of the procedure, as it has never been subjected to official clinical trials. An outcry by gastroenterologists, though, led to the ban being dropped as quickly as it had been put in place. Now, faecal transplants are temporarily permitted, but only for treating *C. diff*.

Imagine yourself in the position of needing, or indeed wanting, to have a faecal transplant to improve your health. You need a donor, and of course you want the finest faeces available. Perhaps the toilet habits of your other half don't seem like something you fancy taking on yourself. Maybe your closest relatives suffer from a twenty-first-century illness themselves, which excludes them from becoming a donor. Short of sending a group email round your friends asking about their constitutions, or posting on Facebook requesting that anybody who rates the behaviour of their bowels to be above average get in touch, what can you do to procure some poo?

This was the situation facing a friend of MIT PhD student Mark Smith in 2011. After eighteen months suffering with a recurrent *C. diff* infection, he was in a bad way. As a pre-med student, he had known from the outset that faecal transplant was an option if his infection wouldn't yield to antibiotics. And, after three failed treatment courses, he was ready to try the next step. The trouble was, he couldn't find a doctor who was willing to perform the transplant. It wasn't the procedure itself that deterred them. Rather, it was the difficult and expensive process of finding and screening a donor, and preparing the faecal infusion. For Mark Smith, whose PhD research focused on the microbes found in water sources, as well as those in the human body, this delay in getting his friend the treatment he needed seemed unacceptable.

Smith reasoned that emergency care doctors whose patients needed a blood transfusion didn't need to rush around recruiting willing blood donors, collecting their blood, testing it for pathogens and compatibility, and packaging it so that it could be delivered to a patient. All they had to do was make a quick phone call to the blood bank, place an order, and continue caring for their patient. Why should it be any different for doctors whose patients needed a transpoosion instead of a transfusion?

While Smith's friend gave in after his seventh failed round of antibiotics, eventually administering his own faecal transplant at home using the unscreened stool of his room-mate, Smith joined forces with an MIT MBA student, James Burgess. Together, and with the support of Smith's PhD adviser, Professor Eric Alm, they set up Open-Biome; a non-profit stool bank. By taking on the donor recruitment, screening, faecal infusion preparation, and shipping of the samples, OpenBiome means that all patients need to find is a willing doctor with a colonoscope, and $250 to cover the costs of the stool sample itself. One hundred and eighty hospitals in thirty-three states are currently making use of OpenBiome's services, meaning 80 per cent of Americans are within four hours' drive of safe, frozen poo, should they need it. Around 2,000 people with *C. diff* have already been treated thanks to OpenBiome's work.

The screening process that volunteers hoping to donate to Open-Biome undergo (it's worth $40 a stool and the heart-warming thought that you could be saving two or three lives with each donation) is nothing that you wouldn't expect: no recent antibiotic exposure or foreign travel, no microbiota-connected health problems, including having allergies or an autoimmune disease, or having metabolic syndrome or major depressive disorder, and no worrying microbes like HIV or *E. coli* 0157. But try finding someone who fits that description. At OpenBiome, to get a single eligible donor, they must quiz and test as many as fifty applicants. Voluntary blood donors, for comparison, are accepted more than 90 per cent of the time.

Although the Centre for Digestive Diseases is a gastroenterology clinic, Borody has seen a handful of remarkable recoveries from illnesses beyond the gut. Not surprisingly, some of his patients with constipation and diarrhoea also suffer from twenty-first-century illnesses. For one man, Bill, who had been suffering from multiple sclerosis (MS) for many years and was unable to walk, a visit to Borody for a faecal transplant was only meant to ease his chronic constipation. But after several days of faecal infusions, Bill started to feel different. As time went by, he regained his health and the ability to walk. Now, it's as if he was never affected by MS.

Bill wasn't Borody's only autoimmune patient who found themselves healthy again after receiving faecal transplants. Two other multiple sclerosis patients, one young woman in the early stages of rheumatoid arthritis, a Parkinson's disease sufferer, and one patient suffering from 'idiopathic thrombocytopenia purpura' – a disease in which the immune system destroys the blood's platelets – have all recovered following faecal transplants. It remains to be seen whether these apparently miraculous recoveries are down to the transplants themselves, or are spontaneous remissions.

Because faeces are not drugs, and all you need is a kitchen blender, some saline and a sieve, with a little help from YouTube videos, anyone can administer their own faecal transplant, and many thousands do. Among those giving it a go, not surprisingly, are the parents of autistic children. Dr Borody himself has seen improvements in autistic children following both faecal transplants and after repeatedly delivering faecal microbes via a flavoured drink. His intention was to relieve the gastrointestinal symptoms, not the psychiatric ones, but Borody says several of the children improved following their treatment. The most encouraging was a young child with a vocabulary of just over twenty words, which shot up to around 800 in the weeks after the microbial therapy. For now, all this is anecdotal. As yet not a single clinical trial has been carried out to test the effects of faecal transplant on autistic patients, though some are planned. The lack of evidence won't stop the parents though – for many, anything is worth a try.

For conditions such as autism and type 1 diabetes, faecal transplants may be – like probiotics – too little, too late. If the damage has already been done, the developmental windows closed, restoring a healthy microbiota may simply halt further damage. For other conditions, with progressively worsening symptoms, it may be possible to turn back the clock.

Remember the experiment in chapter 2 where the gut microbiota from an obese person was put into a germ-free mouse? Two weeks later the mouse had become fat, despite eating no extra food. Well then, what happens if you carry out the experiment in reverse? What happens if you put the gut microbiota from a lean, healthy person into an obese one? For once, I'm not going to tell you the answer to that question in *mice*. But I can tell you what happens if you put these 'lean' microbes into obese *humans*, because that's exactly what a team including the Dutch scientists Anne Vrieze and Max Nieuwdorp from the Academic Medical Center in Amsterdam did.

The aim was not to see whether obese people would lose weight; rather, they wanted to know what immediate impact the lean microbiotas would have. Would receiving a microbial colony from a lean person improve the metabolisms of obese people? I mentioned in chapter 2 the existence of two separate groups of obese people – the healthy and the unhealthy obese. The unhealthy obese – by far the majority – are not merely fat, but also ill. They have symptoms of the most economically important health condition you've probably never heard of: metabolic syndrome. This syndrome is made up of a dangerous constellation of conditions: not just obesity, but type 2 diabetes, high blood pressure and high cholesterol. It costs tens of billions of dollars a year to treat those suffering from metabolic syndrome, and ultimately it is behind most deaths in the developed world. Topping the list of causes of metabolic syndrome-associated deaths are heart disease, obesity-related cancers and strokes.

One aspect of metabolic syndrome – type 2 diabetes – provides a great indicator of a person's health. Unlike in diabetics of the 'type 1' variety, whose insulin-producing cells have been destroyed by auto-immunity, type 2 diabetics can usually still make insulin. The trouble is, their cells don't respond to it. Insulin is a hormone that tells the body's cells to store glucose (sugar) in the blood as fat when it is not needed immediately for energy. Incoming glucose from food is matched by a release of insulin to keep blood glucose levels from becoming perilously high. But if levels of insulin in the blood are per-petually high, the body begins to ignore the request to store glucose. This is insulin resistance, and it's dangerous. Around 30–40 per cent of overweight and obese people have type 2 diabetes. Eighty per cent of them will ultimately die of heart disease.

In healthy people, overweight or otherwise, the blood sugar level spikes after a meal. Blood glucose concentration goes up, insulin is released, and blood glucose falls again. This quick response means that their cells are 'sensitive' to insulin – they are paying attention to the instruction to store the glucose for later use. Amongst people with insulin resistance, there is no blood sugar spike. Blood sugar rises, but comes down only very slowly. Finding a way to reverse insulin resistance could mean preventing deaths associated with metabolic syndrome.

As giving mice microbes from obese people could make them gain weight, Vrieze and Nieuwdorp wondered if giving obese people faecal transplants using the faeces of healthy lean people might im-prove their obesity-related symptoms. They began a clinical trial that they called 'FATLOSE' – Faecal Administration To LOSE insulin resistance. Specifically, could a faecal transplant of a 'lean' microbiota make them more sensitive to insulin? How quickly would their cells store glucose? They treated nine obese men with an infusion made up of stools collected from lean donors, and another nine obese men with an infusion of their own stools as a control. Remarkably, six weeks after their faecal transplants, the men who received the 'lean' microbiotas did indeed become more sensitive to insulin. Their cells were storing glucose nearly twice as quickly as they had been before, nearly matching the insulin sensitivity of the lean healthy donors. The obese men who had had their own stool put back into their colons stored glucose at exactly the same rate after the transplant as before – their insulin sensitivity remained as poor as it had ever been.

It's amazing to think that simply having a different community of microbes living in your gut can make the difference between being healthy and having a condition that carries an 80 per cent chance of death from heart disease. The diversity of the microbiotas of the newly insulin-sensitive men rose from an average of 178 species to one of 234 species. Among these extra species were groups of bacteria that produce the short-chain fatty acid butyrate. Butyrate is thought to play an important part in preventing obesity. The cells of the large intestine are powered by butyrate, preventing the gut from becoming leaky by tightening the protein chains that bind the cells together, and coating themselves with a thicker layer of mucus.

Of course, Vrieze and Nieuwdorp want to know if faecal transplants of lean microbes can also cause weight loss in obese people. They certainly can in mice – giving an obese mouse lean microbes causes a 30 per cent decrease in body fat. A second trial – FATLOSE-2 – is under way to find out if it works in humans too. The results could fundamentally change the treatment of obesity and metabolic syndrome, saving money and improving lives.

If the elements of metabolic syndrome, including type 2 diabetes, can be reversed by restoring a healthy gut microbiota, are faecal transplants actually necessary? Would probiotics have the same effect? Two well-designed trials of *Lactobacillus* species have had encouraging results, benefiting both insulin sensitivity scores and body weight.

Ultimately though, whatever the species, and whatever the outcome, probiotics are a balm. A salve. They pass through us, but they don't hang around for long. To reap their benefits, you must keep taking them. And even if you do take them daily, adding probiotics alone is like sending a foot soldier to war without a box of ammunition.

For a persistent effect, we need to provide an environment where beneficial microbes thrive, day after day, without needing outside intervention to replenish their numbers. And so we come to *prebiotics*. These are not live bacteria, but bacteria food, designed to enhance entire populations of the very healthiest strains. With names like fructo-oligosaccharides, inulin and galacto-oligosaccharides, they sound suspiciously like chemical additives listed on the back of a not particularly natural ready meal. But although they are chemicals, just like any food – from carrots (β-carotene, glutamic acid and hemicellulose, for example) to beef (dimethylpyrazine, 3-hydroxy-2-butanone and many more) – they are not synthetic in origin. They are found in plant foods – the non-digestible fibre we should be eating anyway. But of course, prebiotics are available as food supplements – why eat plants when you can simply sprinkle prebiotic powder on your hamburger?

The benefits of prebiotics, whether isolated or in their original delicious packaging (particularly onions, garlic, leeks, asparagus and bananas, to name a few), might be more wide-ranging than those of probiotics. They have been found to be effective in promoting recovery from food poisoning, in treating eczema, and possibly in preventing colon cancer, but research is still in the early stages. Excitingly, prebiotics may also be a treatment for metabolic syndrome. As I mentioned in chapter 6, they are known to encourage bifidobacteria and *Akkermansia muciniphila*, which help to tighten up a leaky gut, reduce appetite, increase insulin sensitivity, and encourage weight loss.

Ultimately, a faecal transplant is not too different from a probiotic: the idea of both is to deliver beneficial microbes to the gut. One usually goes in at the top and the other at the bottom, and one is usually cultured in a lab whereas the other is cultured in the ideal environment of another person's gut. It's only a matter of time before the two concepts converge. With a well-designed capsule that delivers its contents to exactly the right location in the gut, the same community of faecal microbes that make up the solution used in a faecal transplant can fill a capsule that is swallowed with a glass of water. Its superiority over a probiotic composed of scientific history's leftovers – the lactobacilli – would be huge, with none of the inconvenience, expense and indignity associated with a faecal transplant delivered by colonoscope.

Professor Borody, always on hand with a revolutionary idea and a bit of toilet humour to lighten the mood, is working on such a capsule in collaboration with Alexander Khoruts. Borody calls it the Crapsule. In the relatively relaxed regulatory environment of Australia, he was able to use Crapsule to cure an Australian *C. diff* patient for the first time in December 2014. These pills contain the same unadulterated faecal solution that Borody uses in transplants, but they are delivered orally instead.

The microbiome scientist Dr Emma Allen-Vercoe, however, amidst the tighter regulations of Canada, is working on a more specifically calibrated synthetic stool. Using the same technology that she and her team use in their research into autism – the oxygen-free culture chamber they call Robogut – she and the infectious disease specialist Dr Elaine Petrof, of Queen's University in Kingston, Ontario, have brewed a cocktail of known microbes to use in place of raw stool in faecal transplants. The recipe for this mixture has been refined over forty-one years, in the gut of one supremely healthy woman. Finding someone this healthy took almost as long.

Each year, Allen-Vercoe teaches a class on introductory microbiology to around 300 undergraduate students at the University of Guelph. And each year she asks the group the same question: Has anyone here *never* taken antibiotics? No one ever raises their hand. Like Mark Smith, founder of OpenBiome, Allen-Vercoe was struggling to find anyone, even amongst a young, healthy, athletic population of university students, whose microbes had not been subjected to the potential collateral damage of antibiotic drugs. Eventually, she encountered a woman who had grown up in rural India where antibiotics were far less available. As a child, she'd never had antibiotics, and as an adult, she had only ever received a single dose of them, after having stitches in her knee. She was fit, healthy, disease-free and ate an organic, well-balanced diet. Allen-Vercoe had finally found the ultimate stool donor.

Allen-Vercoe and Petrof used their super-donor's faeces to culture a unique combination of microbes. They selected thirty-three bacterial strains which were not known to be dangerous, were relatively easy to culture, and could be wiped out with antibiotics if necessary. Petrof had two female patients, both of whom had been very ill with recurrent *C. diff* infections for many months, following treatment with antibiotics. Rather than give them a traditional faecal transplant, the plan was to seed the patients' guts with the synthetic stool, which had been named 'RePOOPulate'. Within hours of RePOOPulation, both women were free of diarrhoea, and able to return home. Through the early success of their treatment, Allen-Vercoe and Petrof have jumped the fence from Faecal Microbiota Transplantation to something altogether more modern: Microbial Ecosystem Therapeutics.

While Allen-Vercoe, Petrof and Borody refine their products, and clear the necessary regulatory hurdles, traditional faecal transplants will remain the gold standard in the treatment of recurrent *C. diff* infections. But the future of the faecal transplant is, of course, personalisation. It's all very well using faeces from a screened and scrutinised donor, but why not take it a step further? Think about the process of choosing a sperm donor. Women can watch videos of a prospective biological father being interviewed about his life and his views. They can read a CV of his educational achievements and employment history. They can peruse statistics on height, weight, medical history and the longevity of his genetic stock – biological grandparents- and great-grandparents-to-be. Essentially, in choosing a sperm donor, it is the provenance of the tiny bundles of genes that women are inspecting.

They are choosing these genes to mix with their own – to complement their own. Genes for health and genes for happiness.

The same goes for a poo donor. OK, the genes you get come packaged in microbial wrapping paper, not in a swarm of specialised human cells, but it's genes you're getting, either way. Genes that contribute to height, to weight, to longevity even. Genes that mix with, and complement, your own genes. Genes for health and genes for happiness.

It's inevitable that as faecal transplants become more common, we will, as consumers, demand more from our donors. Although donors are already screened for microbiota-related illnesses, including some mental health conditions, they are not hand-picked or individually matched. Even without going down to the level of the genes, it's easy to imagine the benefits of a little discrimination. What about vegetarian donors for vegetarian recipients, for example? Presumably a microbiota tailored to your diet would make for a smoother handover. Perhaps using a leaner donor might be an added bonus for an overweight recipient in their quest for health. And maybe personality traits could be matched – extroverted microbes, anyone? Or even enhanced – what about a slightly brighter outlook on life courtesy of an optimist's excrement? Perhaps even a side of *Toxoplasma* to spice things up?

This is all fantasy for now, but it is reminiscent of the 'designer baby' debates of the 1990s. Once we get our teeth into the nuts and bolts of microbial genes and exactly how they interact with our own genes, the details of what we're delivering to recipients of faecal transplants might well become a lot more important. At the moment, we're working on the concept that if a gut microbial community is good enough for one person, it's good enough for another. But as we delve into the forces that shape each of our personal colonies – our genetics, our diets, our pasts, our personal interactions, our travels, and so on – there's no doubt we will be more discerning in our selection of the microbes we adopt.

Aiming to restore a healthy microbiota is all very well, but it begs the question: what *is* a healthy microbiota? As Alexander Khoruts, Emma Allen-Vercoe and Mark Smith and his team at OpenBiome have discovered, finding an American who's healthy enough to be considered a worthy stool donor is hard work, with over 90 per cent of well-intentioned volunteers not meeting the screening criteria. When you consider that these people are volunteering on the basis that they see themselves as healthy, the true percentage of Westerners whose microbiotas are worth transplanting probably starts with a zero. So how do antibiotic-scarred, fat- and sugar-riddled, fibre-deprived Western microbiotas compare with the unadulterated microbiotas of people living pre-industrial lifestyles?

Not surprisingly, there's quite a difference. An international team of researchers led by the pied piper of microbiome research, Jeffrey Gordon of Washington University in St Louis, Missouri, collected stools from over 200 people living in two pre-industrial, rural and traditional cultures. The first group came from two Amerindian villages in the Amazonas of Venezuela, where corn and cassava form the basis of a diet that's high in fibre and low in fat and protein. The second group, who eat a similar diet dominated by corn and vegetables, were from four rural communities in the south-east African country of Malawi. Gordon's team sequenced the DNA in each person's faecal microbiota and compared the microbial groups they found with those of over 300 people living in the United States.

Plotting out the microbiotas of the three nationalities according to the degree of similarity among them puts clear water between the US samples and those from the two non-US populations. But the microbiotas of the Amerindians and the Malawians overlapped, with relatively few differences between the microbes they contained. These two groups of people live over 7,000 miles apart, yet their microbiomes are more similar to one another than either is to those of people living in the US. Not only are the US microbiotas distinctly different in their composition, but they are less diverse. Whereas Amerindians averaged over 1,600 different strains of microbes, and Malawians topped 1,400, Americans had fewer than 1,200.

It's hard not to assume that it's the US microbiotas that are the aberration here. By looking at the bacterial groups that differ and the jobs they do, we can judge more easily whether the Western microbiota is damaged or just different. The researchers found that the levels of ninety-two species were especially predictive of whether a microbiota had come from a US or non-US gut. Twenty-three of them belonged to a single genus – *Prevotella*. You might remember this one; it's the same genus that was so very common in the guts of the children from Burkina Faso whom I talked about in chapter 6. The children had this group of bacteria because of their diets – the fibrous plant cells in the grains, beans and vegetables they ate made members of *Prevotella* the dominant species in their guts. Clearly, samples containing these twenty-three *Prevotella* species belonged to non-US microbiotas.

The researchers also took a look at which enzymes were most different between the US and non-US samples. Enzymes are the worker bees of the molecular world – each one has a particular job, breaking apart proteins, for example, or synthesising vitamins. Fifty-two of the enzymes produced by the gut microbes of US and non-US samples really stood out as being different in each population. A quick quiz for you: one group of these discriminatory enzymes were those involved in synthesising vitamins from a vitamin-poor diet – which group, US or non-US, had more of these, do you think? My first thought was that the US sample would need fewer of these, as Western access to nutritious, vitamin-packed foods is surely second-to-none. I had it the wrong way round. In fact, US microbiomes contained *more* genes for vitamin-synthesising enzymes. They also had produced more enzymes for breaking down pharmaceutical drugs, the heavy metal mercury, and the bile salts produced from eating fatty foods.

Essentially, the differences between the US and non-US microbiomes reflected the differences between carnivorous and herbivorous mammals. Where US gut microbes were specialists in breaking down proteins, sugars and sugar substitutes, the Amerindian and Malawian gut microbes were suited to break down the starches found in plants. Food for thought for those considering the palaeo-diet, perhaps.

Microbial restoration is a new and uncertain field of medicine. Whatever the medical potential of probiotics, prebiotics, faecal

world so often do, is not the only marker of health. Even those who live into their eighties and beyond do so within the confines of the quality of life their physical and mental health allows. For toddlers trapped in the torment of autism; for the millions of children suffering with eczema, hay fever, food allergies and asthma; for teenagers being told they must inject themselves with insulin for the rest of their lives; for young adults facing the ruination of their nervous systems; and for the many millions who are overweight, and suffering from depression and anxiety, their quality of life is poorer than it needs to be.

Thankfully, we in the developed world no longer have to take our chances with smallpox, polio or measles, and that very fact represents a great leap forward. But the twenty-first-century illnesses we suffer instead are not a necessary alternative, as the original hygiene hypothesis implied. Our pursuit of a longer life has shifted into the pursuit of a better quality of life for the years that we have. The dogma of the hygiene hypothesis, and its central tenet that infections protect us from allergies and other inflammatory disorders, needs to be pushed aside in the minds of the public and the medical establishment. It is not infections that we're lacking, but Old Friends. We now know that the appendix, once widely assumed to be a pointless vestige of our evolutionary past, is actually a microbial safe-house, providing an education for the body's immune system. Appendicitis, far from being an unavoidable feature of life for at least some of us, is a consequence of the loss of a rich microbial community – of the Old Friends that should protect it from invading pathogens. Rekindling these ancient friendships, the oldest our bodies know, is within our reach.

In chapter 1, I took an epidemiologist's approach to the riddle of the cause of twenty-first-century illnesses, asking where they occurred, who they affected and when they began. The answers reflected changes to the way we live, provided by the wealth and ingenuity of the developed world, where we use antibiotics as medicines for everything from the most benign of colds to the worst of life-threatening infections. Where our farming industries rely on these same drugs to boost the growth of their animals, and to enable huge numbers of genetically similar individuals to be crammed into small spaces and yet resist disease. Where our diets contain the lowest levels of fibre humans have ever eaten. Where so many of our children are not born, but surgically removed from our bodies. And where millennia of mother's milk have been abandoned in favour of formula.

These changes centred on the 1940s, when antibiotics were made available, when diets transformed as the Second World War came to an end, and when both C-sections and bottle-feeding saw a surge in uptake. What has been invisible to us until now is the impact of these changes at a microscopic level. Thousands of generations of co-evolution and cooperation with our symbiotic partners came to an unwitting end the day we declared war on microbes.

Twenty-first-century illnesses affect all of us, from newborns to the elderly, both males and females, of all races. Women bear the brunt of many of them, particularly autoimmune diseases, though it has never been clear why. A neat experiment shows that even this sex difference is connected to the microbiota. In a breed of mice that is genetically prone to getting type 1 diabetes, known as non-obese diabetic mice (NOD), females are twice as likely to develop the disease as males. This sex bias probably has something to do with the impact of hormones on the immune system; castrating male mice makes them more vulnerable, for example. But bring up NOD mice germ-free and the sex difference disappears. The microbiota appear to somehow be controlling the risk of disease. Transferring the microbiota of male mice into female mice protects them from developing diabetes, apparently by raising their testosterone levels. These sex-specific differences are evident only after puberty, though, which explains why type 1 diabetes in humans has no sex bias, as it tends to develop before puberty. In other autoimmune diseases, such as multiple sclerosis and rheumatoid arthritis, the female-to-male bias declines the later in life the condition appears.

From *where*, *who* and *when*, I asked *why* and *how* twenty-first-century illnesses had cropped up. In short, we have damaged our microbiotas. Simply, pushing our microbial communities, in particular those in the gut, out of balance causes inflammation, and inflammation causes chronic disease. We had hoped that the human genome would prove to be a mine of information about the causes of ill-health, but searching amongst our genes has revealed fewer genetically controlled conditions than we had anticipated. Instead, 'genome-wide association studies' (GWAS) have turned up genes that affect only our *predisposition* to different diseases. These gene variants are not necessarily errors, but natural variation that under normal circumstances might not lead to ill-health. When faced with a particular environment, though, genetic differences can make some people more likely than others to develop a particular disease. Tellingly, many of the genetic variants that have been found to be associated with twenty-first-century illnesses are of genes connected to the permeability of the intestinal lining and regulation of the immune system.

In 1900, the top three causes of death in the developed world, claiming the lives of one-third of people, were pneumonia, tuberculosis and infectious diarrhoea. The average lifespan was around forty-seven. In 2005, the top three causes of death, claiming the lives of one-half of people, were heart disease, cancer and stroke. The average lifespan was around seventy-eight. We like to think of these as diseases of old age; an inevitable consequence of living longer lives. But among people living in non-Westernised parts of the world, even those who have run the gauntlet of infectious diseases, accidents and violence, and emerged alive into 'old age', do not tend to die from these three broad categories of disease. What we are now realising is that hearts don't harden, cells don't multiply uncontrollably and blood vessels don't burst just because they are old. The emerging view among medical scientists is that these are not diseases of old age itself, but of inflammation. If there is an effect of advanced age, it is that the modern insults we throw at our bodies have had the necessary time to generate inflammation to the point of catastrophe. If this is the case, a grand old age in the absence of decades of built-up inflammation is a possibility.

Just as the decoding of the human genome heralded a new era

of biology, the recognition of the microbiota as a hidden organ has begun a new era in medicine. Twenty-first-century illnesses have brought fresh challenges for patients living into old age, doctors wanting to provide cures, and pharmaceutical companies designing drugs for chronic consumption. Conventional therapies have reached a dead end when it comes to many of the conditions that afflict us now. We treat long-term, rather than comprehensively cure: antihistamines for allergies, insulin for diabetes, statins for heart disease, and antidepressants for mental illness. The cures for these chronic conditions elude us, because until recently we haven't been able to pinpoint what causes them. Now, with the recognition that the microbiota is not a bystander in the running of our bodies, but an active participant, we have a new opportunity to tackle twenty-first-century illnesses at their source.

So what should we do? Our relationship with our microbes is threatened by three things: our use of antibiotics, the lack of fibre in our diets, and shifts in the way we seed and nurture our babies' microbiotas. We can change each of these both as a society and as individuals.

Societal Changes

The central tenet of medical ethics is 'First, do no harm'. With every treatment comes a risk of unintended side effects, and doctors must balance that risk against the benefits of medication. Until now, the unintended consequences of using antibiotics have been regarded as both minimal and insignificant. In acknowledging the importance of the microbiota to human health, we must also accept that taking a course of antibiotics can sometimes do more harm than good. Even when antibiotics successfully treat an infection, they may cause damage that would ideally be excluded. We already have a compelling reason to reduce our use of antibiotics – the problem of resistance. Despite the risks of this on both a societal and an individual level, it is not proving to be quite worrying enough to persuade doctors and patients to make significant efforts to cut down. But add in the deeply personal consequences of collateral damage to the microbiota, and perhaps we can begin to treat antibiotics the way we treat chemotherapy for cancer – a set of medications with serious consequences for healthy cells, only to be turned to when the benefits outweigh the costs.

There are some more practical steps our society can take both to reduce our reliance on antibiotics, and to reduce their impact when there's no alternative. We know doctors overprescribe these drugs, handing them out to patients even when their illnesses are more than likely to be viral, not bacterial. The trouble is, a doctor cannot usually tell which patients are sick because of a virus, and which because of a bacterium. As things stand, finding out which pathogen is to blame for any given infection involves sending samples to be tested or cultured, and waiting several days for the result. For many patients, and many infections, this is not quick enough. The first step then, in reducing unnecessary antibiotic usage, would be to develop rapid bio-markers that can identify the source of an infection in minutes or hours from easily collected samples, such as stool, urine, blood, or even the breath.

Currently, the broad-spectrum activity of most antibiotics is seen as an advantage. Doctors don't even need to know which bacterial species is responsible for an infection in order to treat it – a broad-spectrum drug will probably be effective. But in an ideal world, we would be able to rapidly identify the bacterium behind an infection, and then treat it using a precisely targeted antibiotic. By looking for molecules specific to each pathogen, we would ideally create antibiotics that destroy them, and only them, sparing the beneficial microbiota that would otherwise become collateral damage. Justifying the extra expense of developing such drugs – one for every pathogen – would rely on a shift away from paying for the consequences of collateral damage down the line, and towards paying upfront for a reduced-risk cure for infections.

Acknowledging the importance of the microbiota does not stop at reducing the use of antibiotics. We could be making use of beneficial microbes as allies in the fight against pathogens. By resisting colonisation by pathogens such as *Staph. aureus*, *C. diff* and *Salmonella*, our resident microbiota do us a great favour. Providing a boost to their defences through improved, targeted and appropriate probiotics could help us to fight infection or to reduce inflammation.

Understanding, and manipulating, an individual's microbiota to improve drug outcomes is the next step in personalised medicine. The heart drug digoxin, for example, needs an individual approach. At the moment, doctors must play a guessing game when choosing what dose of digoxin to prescribe to their patients. For weeks or months, they fine-tune the dosage for each patient they treat, balancing the cost and benefit. The variation in patient response is not down to genetic differences, but to the composition of their gut microbiota. Patients who harbour a single strain of a species of bacterium called *Eggerthella lenta* respond poorly to digoxin, because this common gut microbe inactivates the drug, rendering it ineffective. If cardiologists knew which patients carried *E. lenta*, they could recommend that they increase their protein intake, as the amino acid arginine prevents the bacterium from inactivating digoxin.

Human responses to drugs are far from predictable. Their particular response cannot be predicted by genes and environment alone. The 4.4 million extra genes of the microbiota, part-inherited and part-acquired, play an important part in an individual's response to medicine. Microbes can activate, inactivate and toxify medicines. In 1993, the microbiome's capacity to interfere with medications came at great cost to eighteen Japanese patients, who had developed shingles while suffering from cancer. Their shingles medication had been transformed by their normal gut microbiota into a compound which made their anti-cancer medication lethally toxic. The dangers of the interaction were known when the shingles drug was approved, and the drug's label carried a warning against taking it with anti-cancer medication. Unfortunately at that time it was common practice in Japan for doctors to keep their patients in the dark about a diagnosis of cancer, and to prescribe anti-cancer drugs without a full explanation.

It's easy to imagine that we might begin to sequence the microbes of patients, not only to help in diagnosis, but also to ensure that they receive the most appropriate medication at the right doses. Manipulating the microbiota by adding or removing particular species could help to reduce side effects, improve outcomes and ensure safety. As DNA sequencing continues to reduce in price, the idea that we might monitor the microbiome to assess health risks and chart improvements becomes increasingly realistic.

Our overuse of antibiotics extends to intensive farming. In his excellent book *The Great Food Gamble*, the BBC broadcaster John Humphrys tells the story of visiting a British cattle farm. The farmer proudly showed Humphrys the mighty beasts he had raised using the best drugs that veterinary medicine had to offer. Humphrys noticed a single skinny cow standing in the corner. 'Something wrong with him?' he asked the farmer. 'Nothing at all,' came the reply. 'That's the skinny bugger that's going to end up in our own freezer. The missus doesn't fancy the kids eating all those bloody drugs.'

In the European Union, farmers are no longer allowed to use antibiotics as growth promoters, but inevitably, they are often simply using them as 'treatment' instead. The United States, however, is trailing far behind on banning antibiotic growth promoters, although the Food and Drug Administration has announced its intention to limit their use. It's not only animal products that are affected by antibiotic use, as drug-laced manure can legally be used to fertilise even organic vegetable crops. Ultimately, farming without antibiotics (and those pesticides, hormones and other medications with questionable safety profiles in humans) will cost more, but where would you rather pay, at the checkout, or day-in, day-out through poor health, extra medical costs, and inflated taxes to keep health services afloat?

When it comes to our diets, we are besieged by controversy. Which is worse, butter or oil? How many calories should we eat each day? Are nuts good or bad? Should we cut down on carbs or fat if we want to lose weight? Not even the experts can agree on the answers to these questions, but you'd be hard pushed to find an expert who thinks we shouldn't eat more fibre.

In Britain, the Five-a-Day campaign to get people eating at least five portions of fruit and vegetables each day was launched in 2003. It is so ingrained in the British consciousness that people joke about wine, jam and fruit-flavoured sweets being 'one of my five-a-day'. In Australia, the public health message is 'Go for 2&5' – that's two portions of fruit and five of veg. These messages do seem to have had some impact on people's eating habits, but the trouble is, the focus has been on vitamins and minerals, rather than fibre. As food manufacturers have jumped on the opportunity to get their products noticed, foods as varied as tomato puree and fruit juice have had the seal of approval as counting towards the quota. Fruit, very often juiced or blended, gets more attention than vegetables, and other plant foods such as grains, seeds and nuts are ignored altogether. Fibre, it seems, has not gained the focus it should have. A better message? Eat More Plants.

Perhaps the greatest difficulty we face as a society when it comes to food is the pace of our lives. It is often lack of time that leads to a lack of fibre. A typical lunch on the go in much of the developed world is a sandwich, which might contain a slim layer of salad leaves or a handful of roasted vegetables at best. Hardly a feast of fibre. Dinner too, when there's little time to make it, often consists of hurriedly microwaved ready meals, not renowned for their vegetable content. Even the fruit we eat is packaged for convenience and speed – juiced or blended and bottled – no chopping, peeling or school-bag bruises necessary. Part of the problem is a lack of facilities for cooking and eating in the workplace – vegetables don't taste half as good cold. Even putting a microwave in every workplace would go a long way towards enabling staff to eat meals made with plants. For a culture so focused on food, we put very little effort into eating.

Finally, we come to babies. Over the past century we have made great strides in prenatal care and in reducing infant mortality, especially of babies born prematurely. And, after decades of formula-feeding being more common than breast-feeding, we have, in the developed world at least, come a long way in reinstating breast-milk as the normal food for babies. But in other senses we are going backwards. We place enormous trust in science and medicine, and enormous value on the freedom to choose. As a result, in many cities, Caesarean sections are more common than vaginal deliveries. We should be wary of this intervention: relatively little research goes into women's and children's health issues, particularly in circumstances of 'good health', such as during pregnancy, labour and birth.

Midwives and obstetricians need to know about the microbial consequences of Caesarean sections and formula-feeding. And mothers *deserve* to know. The techniques we use when we bring babies into the world are based on our knowledge of what's best for both mother and baby. But that knowledge continues to evolve. The upshot is that C-sections are not as safe as we thought they were: the altered microbial community that a baby receives may affect their health for days, months and years to come.

The consequence is that we have made an entire generation of children guinea pigs in a huge experiment. What happens if you take babies out surgically, often days or weeks before they're ready to emerge, rather than allowing them to dictate the timing of their appearance, through the appropriately named 'birth canal'? Microbes aside, what happens if you deprive a baby of the hormones released during labour, or of the pressure of being pushed into the world? What happens to a mother's body when she finds herself going from pregnant to not, at the slice of a surgeon's scalpel, rather than after hours of chemical and physical preparation? We are just starting to learn the answers to these questions. As a society, we must reserve Caesarean sections for mothers and babies who really need them, and let nature take its course with those who don't.

The previous generation of children were also guinea pigs in another experiment: what happens when you feed babies the milk of a calf's mother rather than the milk of their own? For around one-quarter of babies born in the developed world today, this experiment persists.

Of course, there are a small number of women (estimates suggest it's fewer than 5 per cent) who are unable to produce the milk needed to satisfy their baby's needs. There are others who have very real difficulties in making breast-feeding work for them and their newborn. Supporting these women and providing good-quality alternatives to mother's milk direct from the breast, whether that's through expressed milk, donated human milk or infant formulas, is crucial. Applying our evolving understanding of human milk oligosaccharides and the microbes supplied in breast-milk to the development of improved infant formulas would make an important contribution to both those mothers who are unable to breast-feed, and those who choose not to.

If midwives, obstetricians and community health-workers are up to date on the latest knowledge of breast-milk, they can offer the best advice and support to parents as they weigh up the options and balance the demands of life on them and their babies.

Individual Changes

In the developed world we're fortunate in that our health is largely dependent on the choices we make. The microbes you are home to, unlike the genes your parents bestowed on you or the infections your environment exposes you to, become yours to shape, cultivate and care for. As an adult, the food you eat and the medicines you take determine the microbes you have. Treat them well and they will return the favour. If you are planning to have children, their microbiotas are up to you as parents, especially if you will be a mother.

I am all for choice. Choice is both a marker and an enabler of freedom. Choice is at the heart of a civilised society. And choice empowers individuals to improve their own lives. But choices made in ignorance of available information are meaningless. The scientific research of the past fifteen years or so into the microbiota has revealed a new layer of complexity and control in the human body. It gives us new insight into how our bodies – as superorganisms – are programmed to function. What choices you make with that information are up to you. All I would suggest is that you make those choices consciously.

I'm advocating that you make a conscious choice about your diet.

Many of my overworked medical doctor friends tell me that one of the greatest frustrations of their day-to-day work is the difficulty of helping patients who won't help themselves. What doctors would often like to prescribe is an active lifestyle and a healthy diet – lower in fat, sugar and salt and higher in fibre. Some patients don't want to hear it; they would rather solve their problems with medicine. But food *is* medicine.

We humans have evolved to be omnivores. Our bodies expect a lot of plants and a little meat. Many of us eat a lot of meat and a few plants, and a whole heap of foods that barely resemble either animals or plants. If you choose to increase your fibre intake by eating more plants, it can be worth doing so slowly and steadily, to give your microbiota time to adapt. An avalanche of fibre into a gut populated by microbes more used to a diet consisting of fats, proteins and simple carbohydrates can produce some undesirable effects. Remember that vegetables, pulses and legumes (that's peas, beans and the like) tend to contain higher fibre contents and lower sugar contents than fruits, and also that blending and juicing can reduce the fibre content and increase access to calories that can be digested by human enzymes and absorbed in the small intestine. If you already suffer from a gastro-intestinal disorder, take the advice of your doctor before changing your diet.

Eating plant foods that encourage a beneficial microbial balance will provide the basis for good health. Make a conscious choice to Eat More Plants.

I'm advocating that you make a conscious choice about your use of antibiotics.

Let me be clear: antibiotics are life-saving drugs, and in many, many situations their benefits outweigh their risks. Yes, we need to consider the microbiota in our decisions about when to take anti-biotics, but without antibiotics we would not have the luxury of caring about our beneficial microbes. The point is not that antibiotics are 'bad' – they are a crucial weapon in our arsenal against pathogenic bacteria – but that we don't need to use cluster bombs to kill spiders.

Doctors are not the only ones who must take responsibility for the overuse of antibiotics. Often, a doctor in general practice will see up to two dozen patients before she even breaks for lunch. With less than ten minutes to listen to a history, make a diagnosis, advise a worried patient and prescribe a suitable drug, a doctor will often give a pushy patient what they want simply so as to be able to move on to their next, equally important, ten-minute appointment. Make a conscious choice about whether *you* want to be that pushy patient.

There are some steps you can take to help you decide whether you need antibiotics. First, consider waiting a day or two to see if your ill-ness gets better. Please note that I said *consider* – use common sense. Second, if your doctor offers you antibiotics, consider asking them the following questions:

1. How sure are you that my infection is bacterial, not viral?
2. Are antibiotics likely to make me significantly better, or help me to recover significantly faster?
3. What are the risks of not taking antibiotics, and allowing my immune system to fight the infection itself?

There is often no clear answer as to which side of the line we should fall on when it comes to using antibiotics, but make a conscious choice, knowing that antibiotics can harm as well as help. Weighing up whether their benefits outweigh their costs is the job of an informed patient in consultation with an informed doctor. Make sure that you are one and that your doctor is the other.

Finally, if treating your illness might involve helping your micro-biota with a good diet, consider doing that. It will always be a good basis for improving your health. Bear in mind that this is an inexact science. The make-up of your microbiota cannot (yet, at least) reveal what illness you might have.

Whether you choose to take antibiotics or not, make your choice consciously.

I'm advocating that you make a conscious choice about giving birth, and feeding your baby.

We are presented with so much information and advice about pregnancy and parenting now that it can sometimes feel as if any natural instincts we might have for the process have been stifled. The good news is that our emerging understanding of the microbiota gives us a straightforward basis for decision-making when it comes to our children: if all is well, stick to nature. If all is not well, Caesarean sections and bottle-feeding are there to help.

The best that any of us can do is be prepared and be aware. Make a conscious choice about giving birth, by creating a birth plan that includes providing your baby with the seeds of a healthy microbiota. The most effective way to do this is by giving birth to your baby vaginally. If you choose to have a C-section, or it's unavoidable, consider adopting the vaginal-swabbing techniques of Maria Gloria Dominguez-Bello. Share your plan with your partner, your doctor and your midwife.

Make a conscious choice about feeding your baby, remembering that breast-milk nurtures the seedlings of the microbiota they receive at birth. If you want to breast-feed, stock up on knowledge, support and determination ahead of time – there's plenty of tips to be found on the internet, including on the website of the World Health Organisation, which provides, alongside information and advice, its recommendations for the optimum duration of breast-feeding for the health and happiness of babies. Don't beat yourself up if it doesn't work out – there are plenty of ways to nurture your child's microbiota.

There's more good news when it comes to child-rearing. We can relax about 'germs'. Most microbes babies encounter in their daily lives will do them no harm. In fact, they'll contribute to a diverse microbiota and help to educate the immune system. Using anti-bacterial sprays and wipes probably does more harm than good.

Whatever choices you make, make them consciously.

In 2000, some of the cleverest members of our species read the DNA code that is responsible for building another four new humans every second of every day. It was a species-defining moment, and one that held particular sway for me as I embarked on the path to becoming a biologist. Looking up at the printed volumes of As, Ts, Cs and Gs that make up our human genome in London's Wellcome Collection years later, I still felt a shiver at the momentous nature of the achievement. That those 120 tomes contained the soul of humanity, and that we had captured its essence on paper, beguiled me, and reinforced my fascination with my chosen career.

It's hard to imagine that the deciphering of the microbiome, even embodied as art in such an iconic way, could have that same magical impact as the decoding of the genome. But our realisation over the past decade or two that the microbes of the human body are part of us, and that their genes are part of our metagenome, may just have an even greater significance for the lives of humans. The microbiota is an organ – the forgotten organ, the unseen organ – within the human body, which contributes to our health and happiness just like any other. But unlike our other organs, this new organ is not fixed. Unlike our human genes, our microbial genes are not unalterable. Both the species we harbour and the genes they contain are our possessions; they are under our control. You can't choose your genes, but you sure can choose your microbes.

Knowledge of our intimate relationship with our microbiotas puts our bodies and our lifestyles into a new context. It is an inescapable connection with our evolutionary past, which ties our modern, technology-filled, nature-deficient, macro-scale lives to their roots. Since Darwin wrote *On the Origin of Species*, we have debated the roles played by nature and nurture in who we are. Is a man tall because his father is tall, or because he grew up with plenty of healthy food?

Is a child clever because her mother is clever or because she had the best teachers? Does a woman get breast cancer because of her genes or because she took synthetic hormones? It's a false dichotomy, of course. Both nature and nurture factor in to the vast majority of traits and diseases. If the Human Genome Project has taught us anything, it's that genes – nature – can *predispose* us to a whole host of conditions, but whether we develop them or not depends on our lifestyle, our diet, and our exposures. In short, our environment: nurture.

Now we have a third player, which sits uncomfortably between nature and nurture. Although the microbiome is strictly an environmental force at work on our eventual characteristics, it *is* genetic, and it *is* inherited. Not via eggs or sperm, not via human genes, but a good portion of the microbiome is passed from parents, especially mothers, to offspring. Many parents hope that they pass on the best of themselves to their children; films like *Gattaca* envisage a future where this wish is not left to chance. Most parents also hope to provide their children with the happiest and healthiest environment that they can manage. The microbiome, with its genetic influence but environmental control, gives parents the power to do both of these things.

Despite all the hype, our human genome did not quite live up to our visions of becoming a blueprint for life, and a philosophy for living. 'It's in our DNA,' we say, as we grapple with our humanity and our idiosyncrasies, but really we know that our own DNA offers little instruction for our daily lives. That other 90 per cent, though – those other 100 trillion cells, and our other 4.4 million genes – they are us too. We evolved with them, and we can't live without them. For the first time, Darwin's theory of evolution, and our other 90 per cent, are showing us the way to live.

Embracing our microbes, who have journeyed with us for millions of years, is the very first step in valuing who we truly are, and, ultimately, being 100 per cent human.

100% Human

In the winter of 2010, when I was in near-constant pain and struggling to stay awake for more than ten hours a day, I was willing to do anything to get better. The possibility that antibiotics – in plastic beads stashed within my hollowed-out toe bone, in liquid form dripping via a cannula into my blood, and in powder-filled capsules dissolving in my gut – might cure me of the infections that were blighting my life was a dream I barely dared hope for. I will always be grateful that these powerful drugs gave me back my health, and the fullness of my life.

They gave me something else as well: an awareness of the 100 trillion friends that share my body. Learning about their contribution to my health and my happiness has given me an entirely new perspective – both on my own life, and on life in a biological sense: the existence, and coexistence, of living things. At the root of my research for this book was a personal pursuit. I wanted to know if I had unwittingly damaged my health by damaging my community of microbes. Even more, I wanted to know whether I could rebuild a microbiota that might help me improve my health.

The beauty of the microbiota is that, unlike our genes, we have some control over it. When I began my research for this book, I sent a sample of my gut microbes to a citizen science programme called the American Gut Project, based at the laboratory of Professor Rob Knight at the University of Colorado, Boulder. By sequencing sections of my bacterial DNA, they were able to tell me which species I played host to. Though I was pleased to see that I had at least some microbes left after my many courses of antibiotics, I was worried by the predominance of two phyla – the Bacteroidetes and the Firmicutes – at the expense of the other phyla. Diversity, it seemed, was something to aspire to. Samples from people living in both the Amazon rainforest and in rural Malawi, whose lives were lived free of antibiotics and an unhealthy Western diet, and who were born vaginally and breast-fed for years, have far greater diversity than people living in the West. I wondered if I might benefit from a few more bugs, and if eating well might help me get them.

Among the groups detected in my sample, I was fascinated to see that I had an unusually high proportion of a genus called *Sutterella*. During my illness, I had developed a tendency to 'tic' when I was tired – the muscles of my face and neck would twitch involuntarily. It was annoying and a little disturbing. *Sutterella*, I learnt, was also over-represented in people with autism, many of whom suffer from physical tics like mine. Could the extra *Sutterella* be responsible for my tics? I wondered. For now, it's impossible to know for sure, as there's still so much research to be done, but it was certainly thought-provoking. Of course, it's worth remembering that this is a scientific field in its infancy. Exactly what roles specific microbial genes, species and communities play in our health and happiness will take time to establish, and we don't yet have the knowledge to diagnose health problems on the basis of the microbiome.

Before I knew about the microbiota – about *my* microbiota – I put little thought into what I ate. I did not subscribe to the notion that 'you are what you eat', and I was sceptical of the idea that foods could impact on my health and well-being in the short term. I did eat relatively healthily, never really bingeing on fast food or sweet treats, but I wasn't too interested in vegetables, eating just one or two portions a day, and I was blissfully unaware of the meagre fibre content of my daily meals. I have always been lean, and I took that as a sign that my diet was a healthy one. Now, though, I think about food entirely differently. I think not only about the nutrition that my human cells extract, but what my microbial cells get to eat as well. Which groups of microbes will benefit from my meal? What will they convert it into? How will those molecules affect my intestinal permeability? What consequences will this have for how I feel? Have I eaten enough microbe-food to justify eating something purely for pleasure?

I haven't had to change much to accommodate my microbiota – my meals are still largely the same, just with a higher fibre content. My breakfast, for example, has gone from being a bowl of packaged, preserved, sugar-laden, fibre-depleted breakfast cereal to a bowl of rolled oats, wheat and barley mixed with nuts, seeds, fresh berries, live unsweetened natural yogurt and milk. It's tastier, cheaper, and it's a feast for my friendly microbes. I'll still eat a weekend fry-up, but I make sure to have a side of beans and a portion of mushrooms. White rice has become brown rice. Lentils replace pasta from time to time. Dense, nutty, rye bread makes an occasional change from toasted slices of fluffy white bread. And I'll add a quick bowl of microwaved frozen peas or steamed spinach to whatever I have for lunch. By my calculations, I've taken my daily fibre intake from around 15 g a day to around 60 g, and it was surprisingly easy to do. Breakfast alone used to contain just 2 g of fibre and is now packed with 16 g – as much first thing as I ate in an entire day before. Ironically enough, the packaged breakfast cereal I used to eat boldly advertised its high fibre content on the front of the box.

So what did these dietary changes do to my microbiota? I had a second sample sequenced once I'd switched to my more microbe-friendly diet. It may not be very scientific, but it most definitely is satisfying to see my efforts evident in my microbes. Topping the list of changes was a bloom in none other than our friend *Akkermansia*, whose association with leanness came up in chapter 2. My second sample contained sixty times the amount of this species than I had had before I embarked on my fibre-based diet. I can just imagine it, gently coaxing my gut lining to produce a nice thick mucus layer, protecting my body from being invaded by the LPS molecules that bother the immune system and alter energy regulation.

The butyrate-producing team of *Faecalibacterium* and *Bifidobacterium*, too, were far more numerous after my dietary intervention than before. I like to think of them helping the cells of my gut lining stay tightly knit together and calming my immune system. So far, so satisfying, but what about its impact on my health? It feels as if things are improving – my fatigue has eased off and my rashes have cleared up, for now at least. Time will tell whether that's luck, the placebo effect, or a genuine result of eating more fibre, but it's not something I'll be giving up. The changes to my microbiota after dabbling in a high-fibre diet are not permanent, of course; to sustain the microbes they feed indefinitely, I have to keep the fibre content of my meals up where it belongs.

Eating for a beneficial microbiota has a pertinence for me beyond my own health. As I contemplate embarking on motherhood, it strikes me that I have more reasons than ever to look after all my cells, both human and microbial. Assuming that taking antibiotics shifted my microbiota for the worse, I'd rather shift it as far back again as I can before passing it on to my future children. If eating more plants helps me achieve that, then that is the conscious choice I make.

Until recently, I had few convictions when it came to birthing and rearing babies. If anything, I trusted in modern medicine to provide the best care for me and my baby. I still do – but only if things go wrong. If not – if all is well – I'll stick to nature. To the birthing process mammals have used for millions of years, and to the milk that has evolved to be just right for growing a human. I'll still have choices to make, and balances to strike, but I will add my new knowledge about the value of the microbiota to the other factors that weigh in to those decisions. Passing on a microbiota that does not consist of microbes from the skin of my belly and the hands of obstetricians and midwives is now a priority for me. If Caesarean section is my only option, it's my intention to mimic nature, swabbing my baby with vaginal microbes as Maria Gloria Dominguez-Bello suggests. As for breast-feeding, I feel ready to arm myself – and my husband – with as much knowledge, strength and support as possible as we work our way through difficult days of pain, exhaustion and inexperience with a new baby. And I hope to be able to breast-feed according to the World Health Organisation's guidelines – six months of exclusive breast-feeding, with further breast-feeding up to two years of age or beyond. That's my aim. And that's my conscious choice.

Finally, I come to antibiotics. These extraordinary drugs brought me full circle, from a past of infectious disease to a present of twenty-first-century sickness. They gave me back a quality of life I feared I'd never regain, but along the way, they took me into new territory that I'd never experienced before. The lesson is not that antibiotics are bad. It's that antibiotics are precious, imperfect, and come with a cost. Since the last dose of my treatment, I have been fortunate enough not to need to take them. If I, or my children, needed them again – really needed them – I wouldn't hesitate. I would take them alongside probiotics, in the hope of lessening the risk of side effects and collateral damage. But if I had the option to wait and see if my immune system might deal with an infection on its own, that's the conscious choice I would make.

As for me and my microbes, we are slowly rebuilding our relationship. Without antibiotics my life would be very different, but now that I am well again, I know one thing: my microbes must come first. After all, I'm only 10 per cent human.

REFERENCES

The scientific literature on the role of the microbiota in human health, both mental and physical, is expanding at an exponential rate. It is a new field, taking off in earnest only a decade or so ago. As well as a great many face-to-face, phone and email conversations with some of the leading scientists in microbiota science, much of the research underpinning this book comes from primary sources – the peer-reviewed research published in scientific journals. The information contained in the book comes from many hundreds of papers – more than I can list here. I include just a handful of references for the most important and interesting of the studies I have written about in *10% Human*, as well as some more general suggestions of general reading about this burgeoning field.

Introduction

1. International Human Genome Sequencing Consortium (2004). Finishing the euchromatic sequence of the human genome. *Nature* 431: 931–945.

2. Nyholm, S.V. and McFall-Ngai, M.J. (2004). The winnowing: Establishing the squid–*Vibrio* symbiosis. *Nature Reviews Microbiology* 2: 632–642.

3. Bollinger, R.R. *et al.* (2007). Biofilms in the large bowel suggest an apparent function of the human vermiform appendix. *Journal of Theoretical Biology* 249: 826–831.

4. Short, A.R. (1947). The causation of appendicitis. *British Journal of Surgery* 53: 221–223.

5. Barker, D.J.P. (1985). Acute appendicitis and dietary fibre: an alternative hypothesis. *British Medical Journal* 290: 1125–1127.

6. Barker, D.J.P. *et al.* (1988). Acute appendicitis and bathrooms in three samples of British children. *British Medical Journal* 296: 956–958.

7. Janszky, I. *et al.* (2011). Childhood appendectomy, tonsillectomy, and risk for premature acute myocardial infarction – a nationwide population-based cohort study. *European Heart Journal* 32: 2290–2296.

8. Sanders, N.L. *et al.* (2013). Appendectomy and *Clostridium difficile* colitis: Relationships revealed by clinical observations and immunology. *World Journal of Gastroenterology* 19: 5607–5614.

9. Bry, L. *et al.* (1996). A model of host-microbial interactions in an open mammalian ecosystem. *Science* 273: 1380–1383.

10. The Human Microbiome Project Consortium (2012). Structure, function and diversity of the healthy human microbiome. *Nature* 486: 207–214.

Chapter 1

1. Gale, E.A.M. (2002). The rise of childhood type 1 diabetes in the 20th century. *Diabetes* 51: 3353–3361.

2. World Health Organisation (2014). Global Health Observatory Data – Overweight and Obesity. Available at: http://www.who.int/gho/ncd/risk_factors/overweight/en/.

3. Centers for Disease Control and Prevention (2014). Prevalence of Autism Spectrum Disorder Among Children Aged 8 Years – Autism and Developmental Disabilities Monitoring Network, 11 Sites, United States, 2010. *MMWR* 63 (No. SS-02): 1–21.

4. Bengmark, S. (2013). Gut microbiota, immune development and function. *Pharmacological Research* 69: 87–113.

5. von Mutius, E. *et al.* (1994) Prevalence of asthma and atopy in two areas of West and East Germany. *American Journal of Respiratory and Critical Care Medicine* 149: 358–364.

6. Aligne, C.A. *et al.* (2000). Risk factors for pediatric asthma: Contributions of poverty, race, and urban residence. *American Journal of Respiratory and Critical Care Medicine* 162: 873–877.

7. Ngo, S.T., Steyn, F.J. and McCombe, P.A. (2014). Gender differences in autoimmune disease. *Frontiers in Neuroendocrinology* 35: 347–369.

8. Krolewski, A.S. *et al.* (1987). Epidemiologic approach to the etiology of type 1 diabetes mellitus and its complications. *The New England Journal of Medicine* 26: 1390–1398.

9. Bach, J.-F. (2002). The effect of infections on susceptibility to autoimmune and allergic diseases. *The New England Journal of Medicine* 347: 911–920.

10. Uramoto, K.M. *et al.* (1999) Trends in the incidence and mortality of systemic lupus erythematosus, 1950–1992. *Arthritis & Rheumatism* 42: 46–50.

11. Alonso, A. and Hernán, M.A. (2008). Temporal trends in the incidence of multiple sclerosis: A systematic review. *Neurology* 71: 129–135.

12. Werner, S. *et al.* (2002). The incidence of atopic dermatitis in school entrants is associated with individual lifestyle factors but not with local environmental factors in Hannover, Germany. *British Journal of Dermatology* 147: 95–104.

Chapter 2

1. Bairlein, F. (2002). How to get fat: nutritional mechanisms of seasonal fat accumulation in migratory songbirds. *Naturwissenschaften* 89: 1–10.

2. Heini, A.F. and Weinsier, R.L. (1997). Divergent trends in obesity and fat intake patterns: The American paradox. *American Journal of Medicine* 102: 259–264.

3. Silventoinen, K. *et al.* (2004). Trends in obesity and energy supply in the WHO MONICA Project. *International Journal of Obesity* 28: 710–718.

4. Troiano, R.P. *et al.* (2000). Energy and fat intakes of children and adolescents in the United States: data from the National Health and Nutrition Examination Surveys. *American Journal of Clinical Nutrition* 72: 1343s–1353s.

5. Prentice, A.M. and Jebb, S.A. (1995). Obesity in Britain: Gluttony or sloth? *British Journal of Medicine* 311: 437–439.

6. Westerterp, K.R. and Speakman, J.R. (2008). Physical activity energy expenditure has not declined since the 1980s and matches energy

expenditures of wild mammals. *International Journal of Obesity* 32: 1256–1263.

7. World Health Organisation (2014). Global Health Observatory Data – Overweight and Obesity. Available at: http://www.who.int/gho/ncd/risk_factors/overweight/en/.

8. Speliotes, E.K. *et al.* (2010). Association analyses of 249,796 individuals reveal 18 new loci associated with body mass index. *Nature Genetics* 42: 937–948.

9. Marshall, J.K. *et al.* (2010). Eight year prognosis of postinfectious irritable bowel syndrome following waterborne bacterial dysentery. *Gut* 59: 605–611.

10. Gwee, K.-A. (2005). Irritable bowel syndrome in developing countries – a disorder of civilization or colonization? *Neurogastroenterology and Motility* 17: 317–324.

11. Collins, S.M. (2014). A role for the gut microbiota in IBS. *Nature Reviews Gastroenterology and Hepatology* 11: 497–505.

12. Jeffery, I.B. *et al.* (2012). An irritable bowel syndrome subtype defined by species-specific alterations in faecal microbiota. *Gut* 61: 997–1006.

13. Bäckhed, F. *et al.* (2004). The gut microbiota as an environmental factor that regulates fat storage. *Proceedings of the National Academy of Sciences* 101: 15718–15723.

14. Ley, R.E. *et al.* (2005). Obesity alters gut microbial ecology. *Proceedings of the National Academy of Sciences* 102: 11070–11075.

15. Turnbaugh, P.J. *et al.* (2006). An obesity-associated gut microbiome with increased capacity for energy harvest. *Nature* 444: 1027–1031.

16. Centers for Disease Control (2014). Obesity Prevalence Maps. Available at: http://www.cdc.gov/obesity/data/prevalence-maps.html.

17. Gallos, L.K. *et al.* (2012). Collective behavior in the spatial spreading of obesity. *Scientific Reports* 2: no. 454.

18. Christakis, N.A. and Fowler, J.H. (2007). The spread of obesity in a large social network over 32 years. *The New England Journal of Medicine* 357: 370–379.

19. Dhurandhar, N.V. *et al.* (1997). Association of adenovirus infection with human obesity. *Obesity Research* 5: 464–469.

20. Atkinson, R.L. *et al.* (2005). Human adenovirus-36 is associated with

increased body weight and paradoxical reduction of serum lipids. *International Journal of Obesity* 29: 281–286.

21. Everard, A. *et al.* (2013). Cross-talk between *Akkermansia muciniphila* and intestinal epithelium controls diet-induced obesity. *Proceedings of the National Academy of Sciences* 110: 9066–9071.

22. Liou, A.P. *et al.* (2013). Conserved shifts in the gut microbiota due to gastric bypass reduce host weight and adiposity. *Science Translational Medicine* 5: 1–11.

Chapter 3

1. Sessions, S.K. and Ruth, S.B. (1990). Explanation for naturally occurring supernumerary limbs in amphibians. *Journal of Experimental Biology* 254: 38–47.

2. Andersen, S.B. *et al.* (2009). The life of a dead ant: The expression of an adaptive extended phenotype. *The American Naturalist* 174: 424–433.

3. Herrera, C. *et al.* (2001). Maladie de Whipple: Tableau psychiatrique inaugural. *Revue Médicale de Liège* 56: 676–680.

4. Kanner, L. (1943). Autistic disturbances of affective contact. *Nervous Child* 2: 217–250.

5. Centers for Disease Control and Prevention (2014). Prevalence of Autism Spectrum Disorder Among Children Aged 8 Years – Autism and Developmental Disabilities Monitoring Network, 11 Sites, United States, 2010. *MMWR* 63 (No. SS-02): 1–21.

6. Bolte, E.R. (1998). Autism and *Clostridium tetani*. *Medical Hypotheses* 51: 133–144.

7. Sandler, R.H. *et al.* (2000). Short-term benefit from oral vancomycin treatment of regressive-onset autism. *Journal of Child Neurology* 15: 429–435.

8. Sudo, N., Chida, Y. *et al.* (2004). Postnatal microbial colonization programs the hypothalamic–pituitary–adrenal system for stress response in mice. *Journal of Physiology* 558: 263–275.

9. Finegold, S.M. *et al.* (2002). Gastrointestinal microflora studies in late-onset autism. *Clinical Infectious Diseases* 35 (Suppl 1): S6–S16.

10. Flegr, J. (2007). Effects of *Toxoplasma* on human behavior. *Schizophrenia Bulletin* 33: 757–760.

REFERENCES

11. Torrey, E.F. and Yolken, R.H. (2003). *Toxoplasma gondii* and schizophrenia. *Emerging Infectious Diseases* 9: 1375–1380.

12. Brynska, A., Tomaszewicz-Libudzic, E. and Wolanczyk, T. (2001). Obsessive-compulsive disorder and acquired toxoplasmosis in two children. *European Child and Adolescent Psychiatry* 10: 200–204.

13. Cryan, J.F. and Dinan, T.G. (2012). Mind-altering microorganisms: the impact of the gut microbiota on brain and behaviour. *Nature Reviews Neuroscience* 13: 701–712.

14. Bercik, P. *et al.* (2011). The intestinal microbiota affect central levels of brain-derived neurotropic factor and behavior in mice. *Gastroenterology* 141: 599–609.

15. Voigt, C.C., Caspers, B. and Speck, S. (2005). Bats, bacteria and bat smell: Sex-specific diversity of microbes in a sexually-selected scent organ. *Journal of Mammalogy* 86: 745–749.

16. Sharon, G. *et al.* (2010). Commensal bacteria play a role in mating preference of *Drosophila melanogaster*. *Proceedings of the National Academy of Sciences* 107: 20051–20056.

17. Wedekind, C. *et al.* (1995). MHC-dependent mate preferences in humans. *Proceedings of the Royal Society B* 260: 245–249.

18. Montiel-Castro, A.J. *et al.* (2013). The microbiota–gut–brain axis: neurobehavioral correlates, health and sociality. *Frontiers in Integrative Neuroscience* 7: 1–16.

19. Dinan, T.G. and Cryan, J.F. (2013). Melancholic microbes: a link between gut microbiota and depression? *Neurogastroenterology & Motility* 25: 713–719.

20. Khansari, P.S. and Sperlagh, B. (2012). Inflammation in neurological and psychiatric diseases. *Inflammopharmacology* 20: 103–107.

21. Hornig, M. (2013). The role of microbes and autoimmunity in the pathogenesis of neuropsychiatric illness. *Current Opinion in Rheumatology* 25: 488–495.

22. MacFabe, D.F. *et al.* (2007). Neurobiological effects of intraventricular propionic acid in rats: Possible role of short chain fatty acids on the pathogenesis and characteristics of autism spectrum disorders. *Behavioural Brain Research* 176: 149–169.

Chapter 4

1. Strachan, D.P. (1989). Hay fever, hygiene, and household size. *British Medical Journal*, 299: 1259–1260.

2. Rook, G.A.W. (2010). 99th Dahlem Conference on Infection, Inflammation and Chronic Inflammatory Disorders: Darwinian medicine and the 'hygiene' or 'old friends' hypothesis. *Clinical & Experimental Immunology* 160: 70–79.

3. Zilber-Rosenberg, I. and Rosenberg, E. (2008). Role of microorganisms in the evolution of animals and plants: the hologenome theory of evolution. *FEMS Microbiology Reviews* 32: 723–735.

4. Williamson, A.P. *et al.* (1977). A special report: Four-year study of a boy with combined immune deficiency maintained in strict reverse isolation from birth. *Pediatric Research* 11: 63–64.

5. Sprinz, H. *et al.* (1961). The response of the germ-free guinea pig to oral bacterial challenge with *Escherichia coli* and *Shigella flexneri*. *American Journal of Pathology* 39: 681–695.

6. Wold, A.E. (1998). The hygiene hypothesis revised: is the rising frequency of allergy due to changes in the intestinal flora? *Allergy* 53 (s46): 20–25.

7. Sakaguchi, S. *et al.* (2008). Regulatory T cells and immune tolerance. *Cell* 133: 775–787.

8. Östman, S. *et al.* (2006). Impaired regulatory T cell function in germ-free mice. *European Journal of Immunology* 36: 2336–2346.

9. Mazmanian, S.K. and Kasper, D.L. (2006). The love–hate relationship between bacterial polysaccharides and the host immune system. *Nature Reviews Immunology* 6: 849–858.

10. Miller, M.B. *et al.* (2002). Parallel quorum sensing systems converge to regulate virulence in *Vibrio cholerae*. *Cell* 110: 303–314.

11. Fasano, A. (2011). Zonulin and its regulation of intestinal barrier function: The biological door to inflammation, autoimmunity, and cancer. *Physiological Review* 91: 151–175.

12. Fasano, A. *et al.* (2000). Zonulin, a newly discovered modulator of intestinal permeability, and its expression in coeliac disease. *The Lancet*, 355: 1518–1519.

13. Maes, M., Kubera, M. and Leunis, J.-C. (2008). The gut–brain barrier

in major depression: Intestinal mucosal dysfunction with an increased translocation of LPS from gram negative enterobacteria (leaky gut) plays a role in the inflammatory pathophysiology of depression. *Neuroendocrinology Letters* 29: 117–124.

14. de Magistris, L. *et al.* (2010). Alterations of the intestinal barrier in patients with autism spectrum disorders and in their first-degree relatives. *Journal of Pediatric Gastroenterology and Nutrition* 51: 418–424.

15. Grice, E.A. and Segre, J.A. (2011). The skin microbiome. *Nature Reviews Microbiology* 9: 244–253.

16. Farrar, M.D. and Ingham, E. (2004). Acne: Inflammation. *Clinics in Dermatology* 22: 380–384.

17. Kucharzik, T. *et al.* (2006). Recent understanding of IBD pathogenesis: Implications for future therapies. *Inflammatory Bowel Diseases* 12: 1068–1083.

18. Schwabe, R.F. and Jobin, C. (2013). The microbiome and cancer. *Nature Reviews Cancer* 13: 800–812.

Chapter 5

1. Nicholson, J.K., Holmes, E. & Wilson, I.D. (2005). Gut microorganisms, mammalian metabolism and personalized health care. *Nature Reviews Microbiology* 3: 431–438.

2. Sharland, M. (2007). The use of antibacterials in children: a report of the Specialist Advisory Committee on Antimicrobial Resistance (SACAR) Paediatric Subgroup. *Journal of Antimicrobial Chemotherapy* 60 (S1): i15–i26.

3. Gonzales, R. *et al.* (2001). Excessive antibiotic use for acute respiratory infections in the United States. *Clinical Infectious Diseases* 33: 757–762.

4. Dethlefsen, L. *et al.* (2008). The pervasive effects of an antibiotic on the human gut microbiota, as revealed by deep 16S rRNA sequencing. *PLoS Biology* 6: e280.

5. Haight, T.H. and Pierce, W.E. (1955). Effect of prolonged antibiotic administration on the weight of healthy young males. *Journal of Nutrition* 10: 151–161.

6. Million, M. *et al.* (2013). *Lactobacillus reuteri* and *Escherichia coli* in

the human gut microbiota may predict weight gain associated with vancomycin treatment. *Nutrition & Diabetes* 3: e87.

7. Ajslev, T.A. *et al.* (2011). Childhood overweight after establishment of the gut microbiota: the role of delivery mode, pre-pregnancy weight and early administration of antibiotics. *International Journal of Obesity* 35: 522–9.

8. Cho, I. *et al.* (2012). Antibiotics in early life alter the murine colonic microbiome and adiposity. *Nature* 488: 621–626.

9. Cox, L.M. *et al.* (2014). Altering the intestinal microbiota during a critical developmental window has lasting metabolic consequences. *Cell* 158: 705–721.

10. Hu, X., Zhou, Q. and Luo, Y. (2010). Occurrence and source analysis of typical veterinary antibiotics in manure, soil, vegetables and groundwater from organic vegetables bases, northern China. *Environmental Pollution* 158: 2992–2998.

11. Niehus, R.M.A. and Lord, C. (2006). Early medical history of children with autistic spectrum disorders. *Journal of Developmental and Behavioral Pediatrics* 27 (S2): S120–S127.

12. Margolis, D.J., Hoffstad, O. and Biker, W. (2007). Association or lack of association between tetracycline class antibiotics used for acne vulgaris and lupus erythematosus. *British Journal of Dermatology* 157: 540–546.

13. Tan, L. *et al.* (2002). Use of antimicrobial agents in consumer products. *Archives of Dermatology* 138: 1082–1086.

14. Aiello, A.E. *et al.* (2008). Effect of hand hygiene on infectious disease risk in the community setting: A meta-analysis. *American Journal of Public Health* 98: 1372–1381.

15. Bertelsen, R.J. *et al.* (2013). Triclosan exposure and allergic sensitization in Norwegian children. *Allergy* 68: 84–91.

16. Syed, A.K. *et al.* (2014). Triclosan promotes *Staphylococcus aureus* nasal colonization. *mBio* 5: e01015–13.

17. Dale, R.C. *et al.* (2004). Encephalitis lethargica syndrome; 20 new cases and evidence of basal ganglia autoimmunity. *Brain* 127: 21–33.

18. Mell, L.K., Davis, R.L. and Owens, D. (2005). Association between streptococcal infection and obsessive-compulsive disorder, Tourette's syndrome, and tic disorder. *Pediatrics* 116: 56–60.

19. Fredrich, E. *et al.* (2013). Daily battle against body odor: towards the activity of the axillary microbiota. *Trends in Microbiology* 21: 305–312.

20. Whitlock, D.R. and Feelisch, M. (2009). Soil bacteria, nitrite, and the skin. In: Rook, G.A.W. ed. *The Hygiene Hypothesis and Darwinian Medicine*. Birkhäuser Basel, pp. 103–115.

Chapter 6

1. Zhu, L. et al. (2011). Evidence of cellulose metabolism by the giant panda gut microbiome. *Proceedings of the National Academy of Sciences* 108: 17714–17719.

2. De Filippo, C. *et al.* (2010). Impact of diet in shaping gut microbiota revealed by a comparative study in children from Europe and rural Africa. *Proceedings of the National Academy of Sciences* 107: 14691–14696.

3. Ley, R. *et al.* (2006). Human gut microbes associated with obesity. *Nature* 444: 1022–1023.

4. Foster, R. and Lunn, J. (2007). 40th Anniversary Briefing Paper: Food availability and our changing diet. *Nutrition Bulletin* 32: 187–249.

5. Lissner, L. and Heitmann, B.L. (1995). Dietary fat and obesity: evidence from epidemiology. *European Journal of Clinical Nutrition* 49: 79–90.

6. Barclay, A.W. and Brand-Miller, J. (2011). The Australian paradox: A substantial decline in sugars intake over the same timeframe that overweight and obesity have increased. *Nutrients* 3: 491–504.

7. Heini, A.F. and Weinsier, R.L. (1997). Divergent trends in obesity and fat intake patterns: The American paradox. *American Journal of Medicine* 102: 259–264.

8. David, L.A. *et al.* (2014). Diet rapidly and reproducibly alters the human gut microbiome. *Nature* 505: 559–563.

9. Hehemann, J.-H. *et al.* (2010). Transfer of carbohydrate-active enzymes from marine bacteria to Japanese gut microbiota. *Nature* 464: 908–912.

10. Cani, P.D. *et al.* (2007). Metabolic endotoxaemia initiates obesity and insulin resistance. *Diabetes* 56: 1761–1772.

11. Neyrinck, A.M. *et al.* (2011). Prebiotic effects of wheat arabinoxylan related to the increase in bifidobacteria, *Roseburia* and *Bacteroides/Prevotella* in diet-induced obese mice. *PLoS ONE* 6: e20944.

12. Everard, A. *et al.* (2013). Cross-talk between *Akkermansia muciniphila* and intestinal epithelium controls diet-induced obesity. *Proceedings of the National Academy of Sciences* 110: 9066–9071.

13. Maslowski, K.M. (2009). Regulation of inflammatory responses by gut microbiota and chemoattractant receptor GPR43. *Nature* 461: 1282–1286.

14. Brahe, L.K., Astrup, A. and Larsen, L.H. (2013). Is butyrate the link between diet, intestinal microbiota and obesity-related metabolic disorders? *Obesity Reviews* 14: 950–959.

15. Slavin, J. (2005). Dietary fibre and body weight. *Nutrition* 21: 411–418.

16. Liu, S. (2003). Relation between changes in intakes of dietary fibre and grain products and changes in weight and development of obesity among middle-aged women. *American Journal of Clinical Nutrition* 78: 920–927.

17. Wrangham, R. (2010). *Catching Fire: How Cooking Made Us Human*. Profile Books, London.

Chapter 7

1. Funkhouser, L.J. and Bordenstein, S.R. (2013). Mom knows best: The universality of maternal microbial transmission. *PLoS Biology* 11: e10016331.

2. Dominguez-Bello, M.-G. *et al.* (2011). Development of the human gastrointestinal microbiota and insights from high-throughput sequencing. *Gastroenterology* 140: 1713–1719.

3. Se Jin Song, B.S., Dominguez-Bello, M.-G. and Knight, R. (2013). How delivery mode and feeding can shape the bacterial community in the infant gut. *Canadian Medical Association Journal* 185: 373–374.

4. Kozhumannil, K.B., Law, M.R. and Virnig, B.A. (2013). Cesarean delivery rates vary tenfold among US hospitals; reducing variation may address quality and cost issues. *Health Affairs* 32: 527–535.

5. Gibbons, L. *et al.* (2010). The global numbers and costs of additionally needed and unnecessary Caesarean sections performed per year: Overuse as a barrier to universal coverage. *World Health Report Background Paper*, No. 30.

6. Cho, C.E. and Norman, M. (2013). Cesarean section and development

of the immune system in the offspring. *American Journal of Obstetrics & Gynecology* 208:249–254.

7. Schieve, L.A. *et al.* (2014). Population attributable fractions for three perinatal risk factors for autism spectrum disorders, 2002 and 2008 autism and developmental disabilities monitoring network. *Annals of Epidemiology* 24: 260–266.

8. MacDorman, M.F. *et al.* (2006). Infant and neonatal mortality for primary Cesarean and vaginal births to women with 'No indicated risk', United States, 1998–2001 birth cohorts. *Birth* 33: 175–182.

9. Dominguez-Bello, M.-G. *et al.* (2010). Delivery mode shapes the acquisition and structure of the initial microbiota across multiple body habitats in newborns. *Proceedings of the National Academy of Sciences* 107: 11971–11975.

10. McVeagh, P. and Brand-Miller, J. (1997). Human milk oligosaccharides: Only the breast. *Journal of Paediatrics and Child Health* 33: 281–286.

11. Donnet-Hughes, A. (2010). Potential role of the intestinal microbiota of the mother in neonatal immune education. *Proceedings of the Nutrition Society* 69: 407–415.

12. Cabrera-Rubio, R. *et al.* (2012). The human milk microbiome changes over lactation and is shaped by maternal weight and mode of delivery. *American Journal of Clinical Nutrition* 96: 544–551.

13. Stevens, E.E., Patrick, T.E. and Pickler, R. (2009). A history of infant feeding. *The Journal of Perinatal Education* 18: 32–39.

14. Heikkilä, M.P. and Saris, P.E.J. (2003). Inhibition of *Staphylococcus aureus* by the commensal bacteria of human milk. *Journal of Applied Microbiology* 95: 471–478.

15. Chen, A. and Rogan, W.J. *et al.* (2004). Breastfeeding and the risk of postneonatal death in the United States. *Pediatrics* 113: e435–e439.

16. Ip, S. *et al.* (2007). Breastfeeding and maternal and infant health outcomes in developed countries. *Evidence Report/Technology Assessment (Full Report)* 153: 1–186.

17. Division of Nutrition and Physical Activity: Research to Practice Series No. 4: Does breastfeeding reduce the risk of pediatric overweight? Atlanta: Centers for Disease Control and Prevention, 2007.

18. Stuebe, A.S. (2009). The risks of not breastfeeding for mothers and infants. *Reviews in Obstetrics & Gynecology* 2: 222–231.

19. Azad, M.B. *et al.* (2013). Gut microbiota of health Canadian infants: profiles by mode of delivery and infant diet at 4 months. *Canadian Medical Association Journal* 185: 385–394.

20. Palmer, C. *et al.* (2007). Development of the human infant intestinal microbiota. *PLoS Biology* 5: 1556–1573.

21. Yatsunenko, T. *et al.* (2012). Human gut microbiome viewed across age and geography. *Nature* 486: 222–228.

22. Lax, S. *et al.* (2014). Longitudinal analysis of microbial interaction between humans and the indoor environment. *Science* 345: 1048–1051.

23. Gajer, P. *et al.* (2012). Temporal dynamics of the human vaginal microbiota. *Science Translational Medicine* 4: 132ra52.

24. Koren, O. *et al.* (2012). Host remodelling of the gut microbiome and metabolic changes during pregnancy. *Cell* 150: 470–480.

25. Claesson, M.J. *et al.* (2012). Gut microbiota composition correlates with diet and health in the elderly. *Nature* 488: 178–184.

Chapter 8

1. Metchnikoff, E. (1908). *The Prolongation of Life: Optimistic Studies*. G.P. Putnam's Sons, New York.

2. Bested, A.C., Logan, A.C. and Selhub, E.M. (2013). Intestinal microbiota, probiotics and mental health: from Metchnikoff to modern advances: Part I – autointoxication revisited. *Gut Pathogens* 5: 1–16.

3. Hempel, A. *et al.* (2012). Probiotics for the prevention and treatment of antibiotic-associated diarrhea: A systematic review and meta-analysis. *Journal of the American Medical Association* 307: 1959–1969.

4. AlFaleh, K. *et al.* (2011). Probiotics for prevention of necrotizing enterocolitis in preterm infants. *Cochrane Database of Systematic Reviews*, Issue 3.

5. Ringel, Y. and Ringel-Kulka, T. (2011). The rationale and clinical effectiveness of probiotics in irritable bowel syndrome. *Journal of Clinical Gastroenterology* 45(S3): S145–S148.

6. Pelucchi, C. *et al.* (2012). Probiotics supplementation during pregnancy or infancy for the prevention of atopic dermatitis: A meta-analysis. *Epidemiology* 23: 402–414.

7. Calcinaro, F. (2005). Oral probiotic administration induces interleukin-10 production and prevents spontaneous autoimmune diabetes in the non-obese diabetic mouse. *Diabetologia* 48: 1565–75.

8. Goodall, J. (1990). *The Chimpanzees of Gombe: Patterns of Behavior*. Harvard University Press, Cambridge.

9. Fritz, J. *et al.* (1992). The relationship between forage material and levels of coprophagy in captive chimpanzees (*Pan troglodytes*). *Zoo Biology* 11: 313–318.

10. Ridaura, V.K. *et al.* (2013). Gut microbiota from twins discordant for obesity modulate metabolism in mice. *Science* 341: 1079.

11. Smits, L.P. *et al.* (2013). Therapeutic potential of fecal microbiota transplantation. *Gastroenterology* 145: 946–953.

12. Eiseman, B. *et al.* (1958). Fecal enema as an adjunct in the treatment of pseudomembranous enterocolitis. *Surgery* 44: 854–859.

13. Borody, T.J. *et al.* (1989). Bowel-flora alteration: a potential cure for inflammatory bowel disease and irritable bowel syndrome? *The Medical Journal of Australia* 150: 604.

14. Vrieze, A. *et al.* (2012). Transfer of intestinal microbiota from lean donors increases insulin sensitivity in individuals with metabolic syndrome. *Gastroenterology* 143: 913–916.

15. Borody, T.J. and Khoruts, A. (2012). Fecal microbiota transplantation and emerging applications. *Nature Reviews Gastroenterology and Hepatology* 9: 88–96.

16. Delzenne, N.M. *et al.* (2011). Targeting gut microbiota in obesity: effects of prebiotics and probiotics. *Nature Reviews Endocrinology* 7: 639–646.

17. Petrof, E.O. *et al.* (2013). Stool substitute transplant therapy for the eradication of *Clostridium difficile* infection: 'RePOOPulating' the gut. *Microbiome* 1: 3.

18. Yatsunenko, T. *et al.* (2012). Human gut microbiome viewed across age and geography. *Nature* 486: 222–228.

Coda

1. Markle, J.G.M. *et al.* (2013). Sex differences in the gut microbiome drive hormone-dependent regulation of autoimmunity. *Science* 339: 1084–1088.

REFERENCES

2. Franceschi, C. *et al.* (2006). Inflammaging and anti-inflammaging: a systemic perspective on aging and longevity emerged from studies in humans. *Mechanisms of Ageing and Development* 128: 92–105.
3. Haiser, H.J. *et al.* (2013). Predicting and manipulating cardiac drug inactivation by the human gut bacterium *Eggerthella lenta*. *Science* 341: 295–298.

LIST OF ILLUSTRATIONS

First plate section

P. 1: Smallpox sufferer (*Wellcome Library, London*)

P. 2: Fat and lean garden warblers (*Franz Bairlein*)

P. 3: [top] Genetically obese (*ob*/ob) mice (*Jackson Laboratory, photo by Jennifer Torrance*); [bottom] Adult obesity trends in the United States (*BRFSS, Centers for Disease Control*)

P. 4: [top] Ellen, Erin and Andrew Bolte on Christmas Day, 1993 (*Ron Bolte*); [bottom] the Bolte family in 2011 (*Christopher Sumpton*)

P. 5: Ant in Papua New Guinea (*Ulla Lohmann*)

P. 6: Limb abnormalities in American frogs (*Stanley Sessions*)

P. 7: Male greater sac-winged bats (*Elizabeth Clare*)

P. 8: Erin Bolte using Robogut (*Alanna Collen*)

Second plate section

P. 1: 'Bubble Boy' David Vetter (*Michelle Goebel*)

P. 2: [top] Penicillin advert (*National Library of Medicine/Science Photo Library*); [bottom] caeca of mice (*Taconic Biosciences Inc.*)

P. 3: [top] Graph of the fat mass of mice (*Cox et al. (2014). Cell 158: 705. Elsevier*); [bottom] Schematic of microbiota transfer to germ-free mice (*Cox et al. (2014). Cell 158: 705. Elsevier*)

P. 4: [top] Anne Miller with Sir Alexander Fleming (*NYT/Redux/eyevine*); [bottom] supermarket aisle (*Chris Pearsall/Alamy*)

P. 5: [top] Joey eating 'pap' (*Lorraine O'Brien*); [middle] hatching Kudzu bugs (*Joe Eger*); [bottom] Peggy Kan Hai on a surfboard (*Peggy Kan Hai*)

LIST OF ILLUSTRATIONS

ACKNOWLEDGEMENTS

It's my contention that science provides the truest source of great new stories that the world has to offer. The recognition of the role that our 100 trillion microbes play in our health and happiness, and the damage we are unwittingly doing to them, is one such story. The twists and turns of the plot have been, and continue to be, unwrapped and fleshed out by hundreds of scientists, and I owe them a debt of gratitude for providing such a rich and fascinating narrative. I have done my very best to represent their discoveries and insights faithfully, and any errors are my own.

Two scientists who have each made an outstanding contribution to microbiota science have made a similarly outstanding contribution to helping me with my research. Patrice Cani and Alessio Fasano have discussed their work, read mine, and responded to my questions with enthusiasm and detail. Special thanks to Derrick MacFabe, Emma Allen-Vercoe, Ted Dinan, Ruth Ley, Maria Gloria Dominguez-Bello, Nikhil Dhurandhar, Garry Egger, Daniel McDonald, Tony Walters and Alison Stuebe, who all helped enormously, giving up time they barely had to spare. Thanks also to Gita Kasthala, David Margolis, Stuart Levy, Jennie Brand-Miller, Tom Borody, Peter Turnbaugh, Rachel Carmody, Fredrik Bäcked, Paul O'Toole, Lita Proctor, Mark Smith, Lee Rowen, Agnes Wold, Erin Bolte, Eugene Rosenberg, Franz Bairlein, Jasmina Aganovic, Jeremy Nicholson, Alexander Khoruts, Maria Carmen Collado, Richard Atkinson, Richard Sandler, Sam Turvey, Sydney Finegold, William Parker, Curtis Huttenhower and Petra Louis, who read drafts, answered questions and were enthusiastic about *10% Human*.

ACKNOWLEDGEMENTS

Further thanks to the very many research scientists whose work I recounted but who were not mentioned by name. Enormous thanks to Ellen Bolte for many hours of conversation and for sharing hers and Andy's story – Ellen, you are an inspiration. And huge thanks to Peggy Kan Hai for letting me share her story and being such a positive force.

I am extremely grateful to the wonderful team at HarperCollins on both sides of the Atlantic. Arabella Pike and Terry Karten were extremely enthusiastic from the start and understood that *10% Human* was a book about humanity, not (just) about microbes – thank you. Thanks also to Jo Walker, Kate Tolley, Katherine Patrick, Matt Clacher, Joe Zigmond, Katherine Beitner, Steve Cox and Jill Verrillo. Huge thanks to my agent Patrick Walsh, whose encouragement kept me on course, and to the team at Conville & Walsh, particularly Jake Smith-Bosanquet, Alexandra McNicoll, Emma Finn, Carrie Plitt and Henna Silvennoinen, whose emails so often make my week. Thanks to the creators of Scrivener who have somehow made it possible to inhabit one's book as it grows.

Thanks to the Ampthill Writers, particularly Rachel J. Lewis, Emma Riddell and Philip Whiteley for getting me out of the house at least once a month. To my friends, thank you for not taking my absence personally and for persistently checking up on me, and thanks to Professor Watson and Miss Adenine for intellectual stimulation. Jen Crees – my virtual office mate – thank you once again for thought-provoking discussions and positive feedback in the early stages. Thanks to my parents, who have been on my side throughout my episodes of ill-health and who never doubt me, and in particular to my mum for listening to endless run-throughs of possible structures. Thanks especially to my best friend and big brother, Matthew Maltby, the master storyteller, for investing so much time and for continuing to tell me the truth, despite my reaction to it. Finally, thanks to Ben, for his unshakable faith in me and for handling the transition from morning bat facts to evening microbe facts with such good grace.

INDEX

Page numbers in *italic* refer to the illustrations

307